GEOMETRY
OF
CONSTRUCTION

GEOMETRY OF CONSTRUCTION

Second Edition
By T. B. Nicholls and N. P. Keep

With an introduction by Richard Hoyle and Daren Tailby

LONDON AND NEW YORK

First Edition published in 1947 by Cleaver-Hume Press Ltd, London.
Second revised edition published in 1954 by Cleaver-Hume Press Ltd.

Published by Donhead Publishing Ltd 2009

Published 2015 by Routledge
2 Park Square, Milton Park, Abingdon, Oxon OX14 4RN
711 Third Avenue, New York, NY 10017, USA

Routledge is an imprint of the Taylor & Francis Group, an informa business

© Taylor & Francis 2009
New introduction to this edition © Richard Hoyle and Daren Tailby 2009

All rights reserved. No part of this book may be reprinted or reproduced or utilised in any form or by any electronic, mechanical, or other means, now known or hereafter invented, including photocopying and recording, or in any information storage or retrieval system, without permission in writing from the publishers.

Product or corporate names may be trademarks or registered trademarks, and are used only for identification and explanation without intent to infringe.

ISBN 13: 978-1-873394-89-2 (hbk)

British Library Cataloguing in Publication Data

Nichols, Trafalgar Bertram.
  Geometry of construction.
  1. Architectural drawing. 2. Geometrical drawing.
  I. Title II. Keep, Norman.
  720.2'84-dc22

ISBN-13: 9781873394892

Introduction to the 2009 edition

Geometry of Construction, first published in 1947 by Cleaver-Hume Press Ltd of London, remains the most concise and instructive guide to the technical geometry of the construction industry. Specifically written to develop the skills of students and apprentices, whether architects, carpenters, stonemasons, engineers or other crafts, this book covers all aspects of two and three dimensional geometry found in technical drawing and constructional practice.

Trafalgar Bertram Nichols A.I.O.B., M.I.P.H.E., A.I.Struct.E., was born in 1908 at Orsett, Essex. He progressed through industry into teaching and by the early 1930s had become Head of the Building Department at Guildford County Technical College, (now Guildford College of Further and Higher Education). It was during this time that Nichols began to write technical books about the construction industry. The first, *An Introduction to Masonry*, was published in 1936 by English University Press of London, and was closely followed by *Stonework* published in 1938 by Crosby Lockwood of London. Nicholls later wrote *Building Craft Certificate Series*, published in 1961 by Cleaver-Hume Press Ltd of London.

Norman Presto Keep F.R.I.B.A., was born in 1894 at Wandsworth, London. He was an experienced member of RIBA (Royal Institute of British Architects) and taught Construction at Willesden Technical College (now The College of North West London) where he was Head of the Building Department. He also became the chief examiner in Building for Surrey County Council.

When Nichols and Keep collaborated to write *Geometry of Construction*, the combination of their vast practical experience and their lecturing skill produced what many consider to be the best geometry book ever written for the construction industry.

Beginning with the most basic principles in technical drawing and moving on to more complex and intricate details, Nichols and Keep demonstrate their wide experience of architectural drawing. This combines with their teaching skills to give clear, detailed, yet straightforward instructions by which readers can quickly grasp the fundamental principles behind construction geometry. A feature of the book is its layout, a model of clarity, linking concise instructions with extremely clear diagrams.

The contents are delivered to the reader in a progressive series of exercises that will develop their skills as they work through its pages. The more experienced can turn to an appropriate page to be reminded of the method to tackle a particular project.

The first part examines topics such as lettering, bisectors, angles, chords, the circle and the construction of scales. These are essential basics for students before they can advance to drawing skills. It is necessary to master the methods in this area as the following chapter provides examples of projected drawings and mouldings, which are more challenging to draught correctly, demanding a higher level of technique using projection methods. It explains how to use orthographic, isometric, oblique and axonometric drawings, while giving detailed examples of each.

The next part then looks at the circle and how to create mouldings and arch components, using arcs and bisection methods. Once again Nichols and Keep show that much work can be completed purely with the simple tools of dividers, compass and rule.

The following section contains various fret patterns and designs; they are as interesting to draw as they are to produce in stone or timber. Many of the circular designs overlap, so care needs to be taken when carving these sections to prevent inadvertently removing material that will be needed as part of the design. This section also describes the correct detail to create an ellipse, parabola or hyperbola, with the necessary calculations to work out areas and volumes as well. The final part of this section details the entasis of a column and an ionic volute; both are a test for the draughtsman, but are extremely satisfying to draw. The volute is a complex design that requires extreme precision to complete, as the construction lines are very close to one another at its centre. Geometrical solids and projections of solids, provided at this point, are useful examples, since being able to develop shapes and moulds is an integral part of the setting out process in stonemasonry.

The final section in *Geometry of Construction* continues to develop the skills of the student as the drawings become ever more detailed. Examples of this complexity can be seen in the cutting of a sphere, continued developments and interpenetrations of moulds. An excellent example of a pediment moulding shows clearly how setting out should be done. Intricate tracery is also shown in detail; this work is popular with masons as it is a challenge to produce. Another great technical challenge is that of the hemispherical dome and the setting out of barrel vaulting. Careful attention to detail must be paid when drawing, as all sections have to be correct - mistakes can be costly if drawings with errors go through to the production process. The last pages detail shadow projection within a drawing, difficult to produce and a test of highly skilled draughtsmanship.

Geometry of Construction was reprinted in 1948 and this second enlarged edition was produced in 1954. It was written in response to the demand and encouragement from lecturers and students to add further pages of drawings and accompanying text concerning raking sections, the projection of points, of lines, and of planes, the true lengths of lines, the oblique plane and on roof surfaces. These additions were incorporated into each relevant section within the original book. One technical magazine of the time (*Illustrated Carpenter & Builder*) is quoted as saying:

Authors and publishers deserve to be congratulated on the production of an attractive book to suit the needs of students taking National Certificate, L.I.O.B., City & Guilds and similar courses. Arrangement of the text on the left hand pages and the corresponding diagrams on the right hand pages is commendable in a book for students, as it makes each page and facing page a self-contained lesson. Draughtsmanship is of a high standard, while the text is clear, concise and adequate.

Unfortunately after several reprints, this valuable book gradually became unavailable to purchase or even to borrow from a library, and students were denied the benefit of this excellent guide. Now republished by Donhead Publishing, the second edition of *Geometry of Construction* provides again one of the most detailed and valuable resources to develop the knowledge and skills of a draughtsman. It also offers an unparalleled insight into the world of drawing itself.

Richard Hoyle and Daren Tailby
(June 2009)

R.J. Hoyle – Stonemasonry lecturer/Advanced Practitioner
Richard Hoyle was born in Yorkshire to a family whose history in the stone industry now spans three centuries. Richard developed a great passion as a young boy to follow in his ancestors' footsteps and become a skilled stonemason, continuing the family tradition. After leaving school, Richard served his apprenticeship with a local stonemasonry firm and trained at York College under the expert guidance of lecturer Kevin Calpin. With a firm background and knowledge of the industry behind him, he started his own business, which ran successfully for five years, before making the decision to move into lecturing stonemasonry at Moulton College, Northampton, where he is now one of the College's advanced practitioners.

Daren Tailby – Stonemasonry lecturer
With over 25 years experience in the stone industry, Daren Tailby is a highly skilled, experienced and accomplished mason. Starting work at the age of fifteen for a local company in Kettering, Northamptonshire, he quickly developed an extensive range of skills which include letter cutting, carving and banker masonry, these skills being further enhanced by the quality of his draughtsmanship. Daren has worked for several monumental and banker masonry companies, on many prestigious buildings across the country, from churches and cathedrals to stately homes and private dwellings. Keen to share his extensive range of skills and experience, Daren now lectures in stonemasonry at Moulton College.

GEOMETRY OF CONSTRUCTION

By

T. B. NICHOLS
A.I.O.B. M.I.P.H.E. A.I.Struct.E.

*Head of the Building Department,
County Technical College, Guildford*

and

NORMAN KEEP
F.R.I.B.A.

*Head of the Building Department,
Willesden Technical College;
Chief Examiner in Building,
Surrey County Council*

SECOND EDITION
ENLARGED

LONDON
CLEAVER–HUME PRESS LTD.

Cleaver-Hume Press Ltd.
31 Wright's Lane, Kensington
London, W.8

All rights reserved

First published 1947
Second Impression 1948
Second Edition 1954
Second Impression 1956
Third Impression 1959

PRINTED IN GREAT BRITAIN

PREFACE

THE considerable popularity which this book had attained among teachers and students, encouraged the authors to add in this second edition some additional sections for which their colleagues have from time to time expressed a need. There are new pages of drawings, with covering text, on Raking Sections, the Projection of Points, of Lines, and of Planes, the True Lengths of Lines, the Oblique Plane, and on Roof Surfaces.

A knowledge of geometry is of importance to the designer and craftsman alike, since it not only forms the essence of all technical drawing but also the basis for most constructional work. A large number of everyday problems of the workshop and drawing-office may be readily solved by application of its basic principles.

In this volume the authors have endeavoured to give a thorough grounding in geometry and to illustrate its application to building practice.

The work is arranged progressively and is planned to be suitable for use by students studying for National Certificates in Building, for the Licentiateship of the Institute of Builders, and for the City and Guilds of London Institutes examinations. It covers the syllabuses of Secondary Technical Schools and Senior Building and Architectural courses, whilst the needs of students engaged in private study have not been lost sight of.

The arrangement of the text on the left-hand pages and the corresponding diagrams on the right-hand pages eliminates the necessity for referring back, a feature which has been appreciated by technical students who often have to work under conditions which are far from ideal.

The authors are indebted to the late Mr. C. Hetherington for his valuable criticisms during the preparation of the drawings and text, and to Mr. P. J. Edmonds for textual suggestions.

T. B. N.
N. K.

Summer 1959

CONTENTS

	PAGE
Lettering for Working Drawings	8
Some Definitions	12
Bisectors and Perpendiculars	14
Angles by Bisection	16
Angles and Triangles	18
Construction of Triangles	20
Scale of Chords	24
Quadrilaterals and Parallel Lines	26
Construction of Quadrilaterals	28
Regular Polygons and their Construction	30
The Circle	34
The Construction of Scales	36
Enlargement by Squares	38
Orthographic Projection	40
Arrangement of Projections	42
Isometric Projection	44
Oblique Projection	48
Axonometric Projection	50
The Projections of a Circle	52
Orthographic Projection: an Example	54
Isometric Projection: an Example	56
Axonometric Projection: an Example	58
Oblique Projection: an Example	60
The Circle in Mouldings	62
Arch Curves	64
Mouldings	68
Tangents to Circles: Construction	70
Circles in Contact	74
Tangents to Circles: External and Internal	76
Inscribed Circles	78
Foiled Figures	80
Continuous Curves	82
Loci: Locus of Centres	86
Fret Patterns	88
Patterns Based on Squares	90
Patterns Based on Circles	92
Patterns in Circles	94

CONTENTS—*continued*

	PAGE
The Ellipse as a Plane Figure	96
The Parabola and Hyperbola	98
Conic Sections in Mouldings	100
Approximations to Ellipse	102
Areas	104
Calculation of Areas	108
Calculation of Volumes	110
Similitude	112
Entasis of Column	114
Spiral Curves	116
Ionic Volute	118
Geometrical Solids	122
Projections of Solids: Cube	124
do. Triangular Prism	126
do. Hexagonal Prism	128
do. Cylinder	130
do. Square Pyramid	132
do. Triangular Pyramid	134
do. Hexagonal Pyramid	136
do. Cone	138
Auxiliary Projection	140
Sections of Solids	142
The Ellipse as a Conic Section	150
Projection of Points	152
Projections of Lines	154
Projections of Planes	156
The Lengths of Lines	158
The Oblique Plane	160
Roof Surfaces	162
The Parabola as a Conic Section	164
The Hyperbola as a Conic Section	166
Cuttings of Sphere	168
Developments	170
Interpenetration	174
Intersecting Mouldings	188
Pediment Mouldings	190
Raking Sections	192
Tracery	194
Tracery Panels	196

CONTENTS—continued

	PAGE
Geometrical Tracery	198
Developments of Sphere	200
Hemispherical Dome	202
Barrel Vaulting	204
The Helix	210
Shadow Projection	212
Index	227

LETTERING FOR WORKING DRAWINGS

DRAUGHTSMANSHIP is the art of conveying technical directions by means of working drawings. These drawings are invariably amplified by the addition of figured dimensions and descriptive legends, which must be arranged in an orderly manner, clearly printed and decipherable. For these reasons, a few words on lettering generally seem justified.

Various types of lettering are shown here, any one of which may be used where suitable to the particular job in hand.

General purpose lettering used in offices may be divided into two types, sloping and upright. Each office has its own type, but whichever is used legibility must be the first consideration. "Rustic", "Gothic", and other fancy types should be avoided.

Two faint pencil guide lines should be drawn to maintain the level of the letters, and erased on completion.

Main headings should be roughed in with a pencil until the draughtsman has gained experience. In this manner imperfect lettering may be corrected, bunched words respaced, and any excessive spaces filled up.

When inking in, it is important to select a nib suitable to the purpose and to the individual's hand.

However well a drawing is presented it will be ruined by poorly-designed and weakly-executed lettering, therefore the constant practice of lettering will bring its own reward in giving distinction and character to the drawings.

Some firms use stencil lettering in order to save time, but the saving is negligible once the draughtsman has become experienced.

ABCDEFGHIJKLM
ABCDEFGHJKLMN
GVIDE LINES IN PENCIL
PLAN ELEVATION SECTIO
abcdefgghijklmnopqrstu
PLAN AND ELEVATION OF
1234567890
ONIVS·MFB·BVSPROCOVS·
BASILICAMPORTASMVRVMPECVNIA
ARCHITECTVRAL
PERSPECTIVE
abcdefghijklmnop
132 qrstuvwxyz 46
ABCDEFGHIM

LETTERING

LETTERING—continued

The ROMAN LETTER is considered the best for both legibility and decorative effect. The alphabet illustrated here is taken from the well-known inscription on the base of Trajan's Column, Rome (c. A.D. 114) and is the inspiration of all noble lettering. A plaster cast of this inscription, which consists of six rows of capitals incised on a rectangular slab, is in the Victoria and Albert Museum, South Kensington.

This cast should be carefully studied with regard to the ratio between thick and thin strokes, the spacing of words, and the overall sense of beauty that the formation of the letters and lay-out give.

It should be noted that the inscription was cut into the stone and the letters were set out without the use of instruments.

The student's attention is also drawn to the modern memorial stone in the foyer of the R.I.B.A. headquarters, beautifully lettered by Mr. Ernest Gillick and showing the skilful design of connected capital letters.

Many attempts have been made without success to produce a "standard", geometrically constructed, alphabet.

Some guide lines are indicated on the letters shown here to convey an idea of proportion between them. After studying this example in all its aspects of proportion, spacing, and character, students should develop their own style, while avoiding all eccentricities.

SOME DEFINITIONS

A PLANE, Fig. 1, is a flat surface. A line drawn between any number of points in a plane is wholly in that plane. A straight edge placed on the plane and revolved will be in contact with it in every position. A carpenter or mason tests the accuracy of a surface by placing straight edges across either end and sighting beneath them. If the surface is a true plane the straight edges will appear to coincide.

A SURFACE has length and breadth but no thickness. A plane is a surface but a surface is not necessarily a plane. The top side of the tile shown in Fig. 2 is a surface but not a plane.

A SOLID, Fig. 3, has three dimensions: length, breadth, and depth or thickness. A solid is bounded by surfaces, a surface by lines, and a line by points.

PLANE GEOMETRY deals with figures drawn on plane surfaces, and which have therefore no thickness. SOLID GEOMETRY deals with three-dimensional or solid objects.

A STRAIGHT LINE is the shortest distance between two points, it has the same direction throughout its length and has no thickness, Fig. 4. A CURVED LINE, Fig. 5, changes its direction throughout its length. A POINT indicates position but has no magnitude. A point, when represented, must therefore be very small and a line very fine.

An ANGLE is formed by the meeting of two straight lines (termed the arms) in a point (termed the vertex), Fig. 6. Note the lengths of the arms of an angle do not affect its size.

PARALLEL LINES are straight lines in the same plane; when produced they will never meet, Fig. 8.

A TRIANGLE is a plane figure bounded by three straight lines.

Geometrical Construction

Students commencing the study of Practical Geometry should master thoroughly the first simple geometrical constructions given in the following pages. Much future work depends on these, and many constructions which appear to be complicated really consist only of a combination of simple operations. It is essential to dissociate the construction from any particular mind picture of an arrangement of lines. The conditions in practice may be different from those in the text-book; a triangle or rectangle standing in an unusual position may appear as quite another figure unless the fundamental principles have been understood.

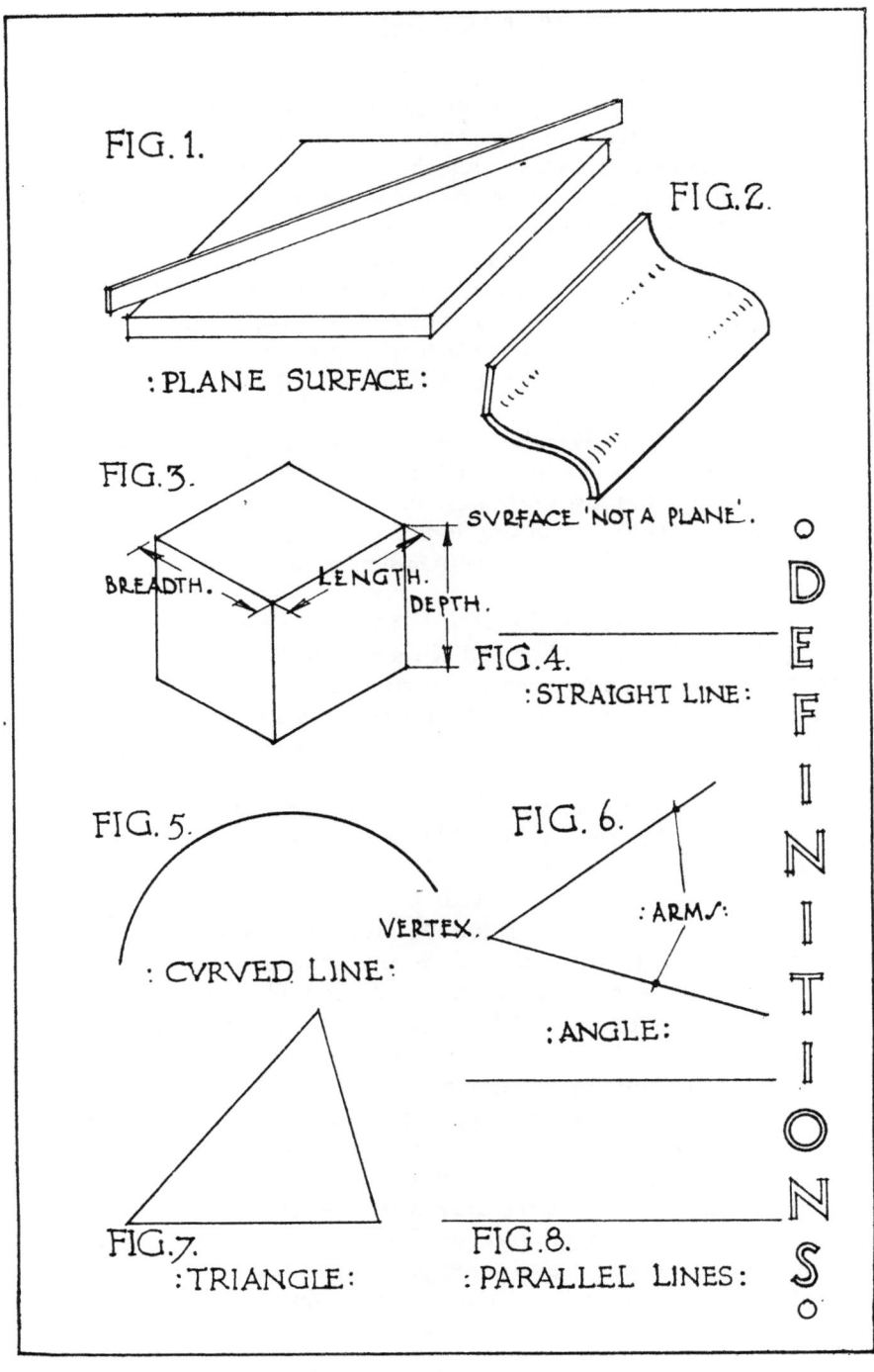

BISECTORS AND PERPENDICULARS

To bisect (i.e. to divide into two equal parts) *a given straight line*, Fig. 1: Let AB be the given straight line. With the extremities of the line as centres, and any radius greater than half the length of the line, draw intersecting arcs above and below the line. A line drawn through these intersections bisects the line in the point X and is perpendicular to the line at that point

To bisect a given angle, Fig. 2: Let ABC be the angle. With centre B and any radius draw arcs cutting the arms of the angle in D and E. Using these points as centres and the same or any other convenient radius draw two further arcs intersecting in point X. Join BX. This is the required bisector; angle ABX is equal to angle CBX.

To erect a perpendicular from a given point in a straight line, Fig. 3: Let AB be the line and X the point. With X as centre and any convenient radius draw arcs cutting the line in C and D. With C and D as centres and any larger radius draw arcs intersecting above the line in E. EX is perpendicular to AB, i.e. angles EXA and EXB are right angles.

To draw a perpendicular to a line from a given point outside it, Fig. 4: Let AB be the line and X the point. With X as centre and any convenient radius draw arcs cutting the line in C and D. With C and D as centres draw arcs intersecting below the line in point Y. YX is perpendicular to AB, but is not necessarily a bisector of the line.

To erect a perpendicular from a given point in a straight line, Fig. 5: Let AB be the line and X the point. (NOTE.—*This is a second method of solving the problem in* Fig. 3.) With centre X and any radius draw a semicircle. With the same radius draw two arcs cutting the semicircle in E and F. With the same or any larger radius, and centres E and F, draw arcs intersecting in G. GX is perpendicular to AB at the point X. When finding points E and F, the radius may be marked from points C and D, as explained, or from D to F and from F to E, because the radius of a circle steps round the circumference six times or round the semicircle three times.

To erect a perpendicular to a given line from a given point situated at or near the end of the line, Fig. 6: Let AB be the given line and X the given point on it. Choose any point O outside the line. With centre O and OX as radius draw a portion of a circle cutting AB at C. Join CO and produce the line until it cuts the arc at D. Join D to X; DX is the required perpendicular.

NOTE.—The angle in a semicircle is a right angle.

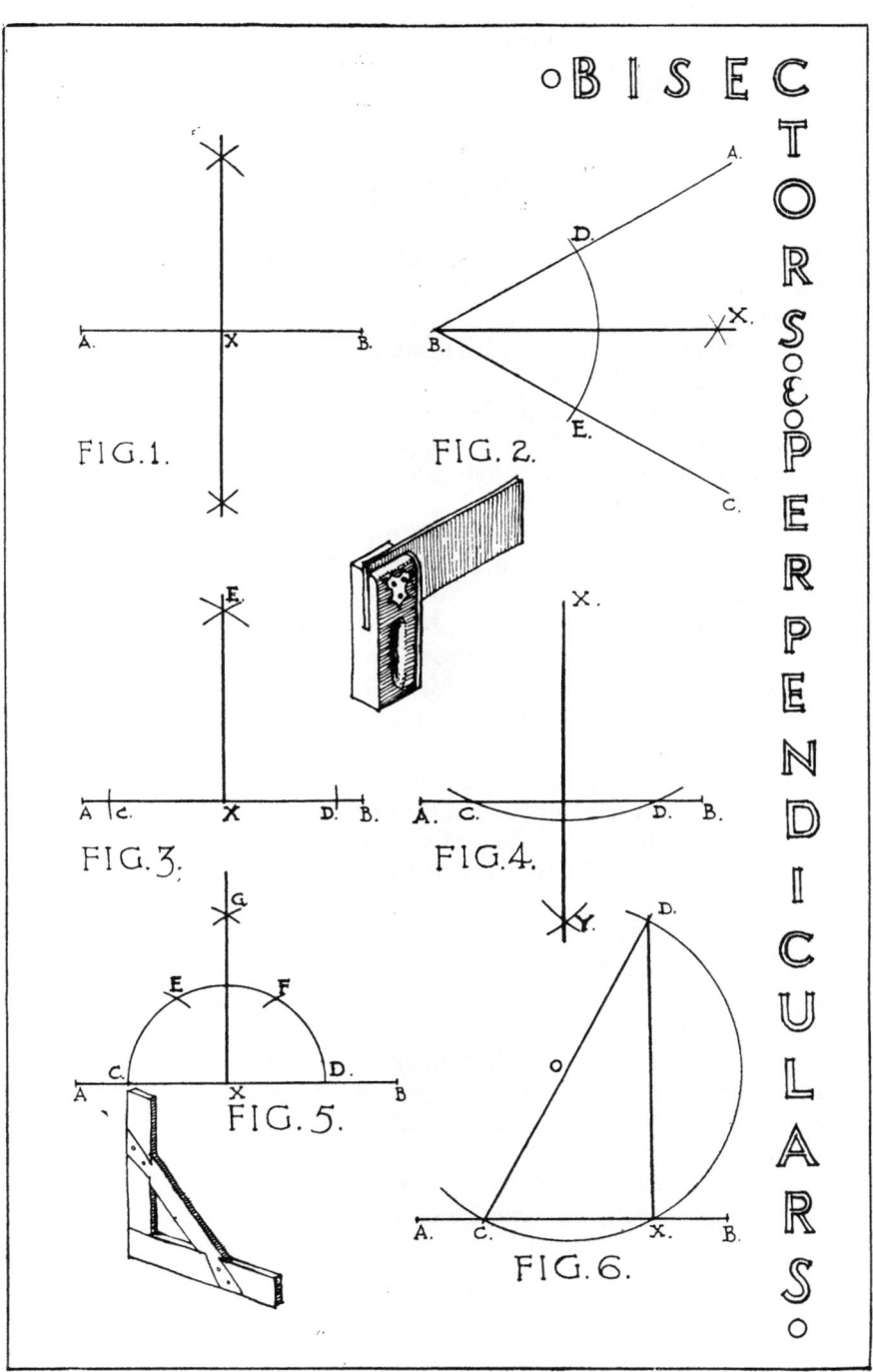

ANGLES BY BISECTION

A COMPLETE revolution consists of 360 degrees (°); each degree may be divided into 60 minutes ('), and each minute into 60 seconds (").

The division of the revolution into a series of angles is shown in Fig. 4, and is as follows:

1. ACUTE ANGLE. An angle less than a right angle.
2. RIGHT ANGLE. 90° or a quarter of a revolution.
3. OBTUSE ANGLE. An angle greater than a right angle but less than two right angles.
4. STRAIGHT ANGLE. Two right angles or a straight line.
5. REFLEX ANGLE. An angle greater than two right angles but less than four right angles.

It is usual in drawing-office practice to obtain angles by the use of either set squares or the protractor, the adjustable set square being particularly useful. Most of the angles met with in construction can, however, be obtained geometrically by the bisection of others.

The radius of a circle steps round the circumference six times, therefore if an arc is struck, and a portion of it cut off, using the same radius, the angle subtended by that arc at the centre equals $\frac{1}{6}$ of 360° or 60°. The bisector of 60° gives 30° and the bisector of 30° gives 15°, Fig. 1.

Fig. 2 shows the construction of angles of 90°, 45°, and $22\frac{1}{2}$°.

The right angle is constructed by any of the methods previously explained: 45° by bisecting 90°, and $22\frac{1}{2}$° by bisecting 45°.

Fig. 3 shows obtuse angles of 120°, 135°, and 150°. They are set out by first constructing their *supplements*. Supplementary angles are angles that together total two right angles or 180°. It will be seen that 180° − 60° = 120°; 180° − 45° = 135°; 180° − 30° = 150°.

It will be readily seen that a considerable number of angles in addition to those shown can be obtained by these methods.

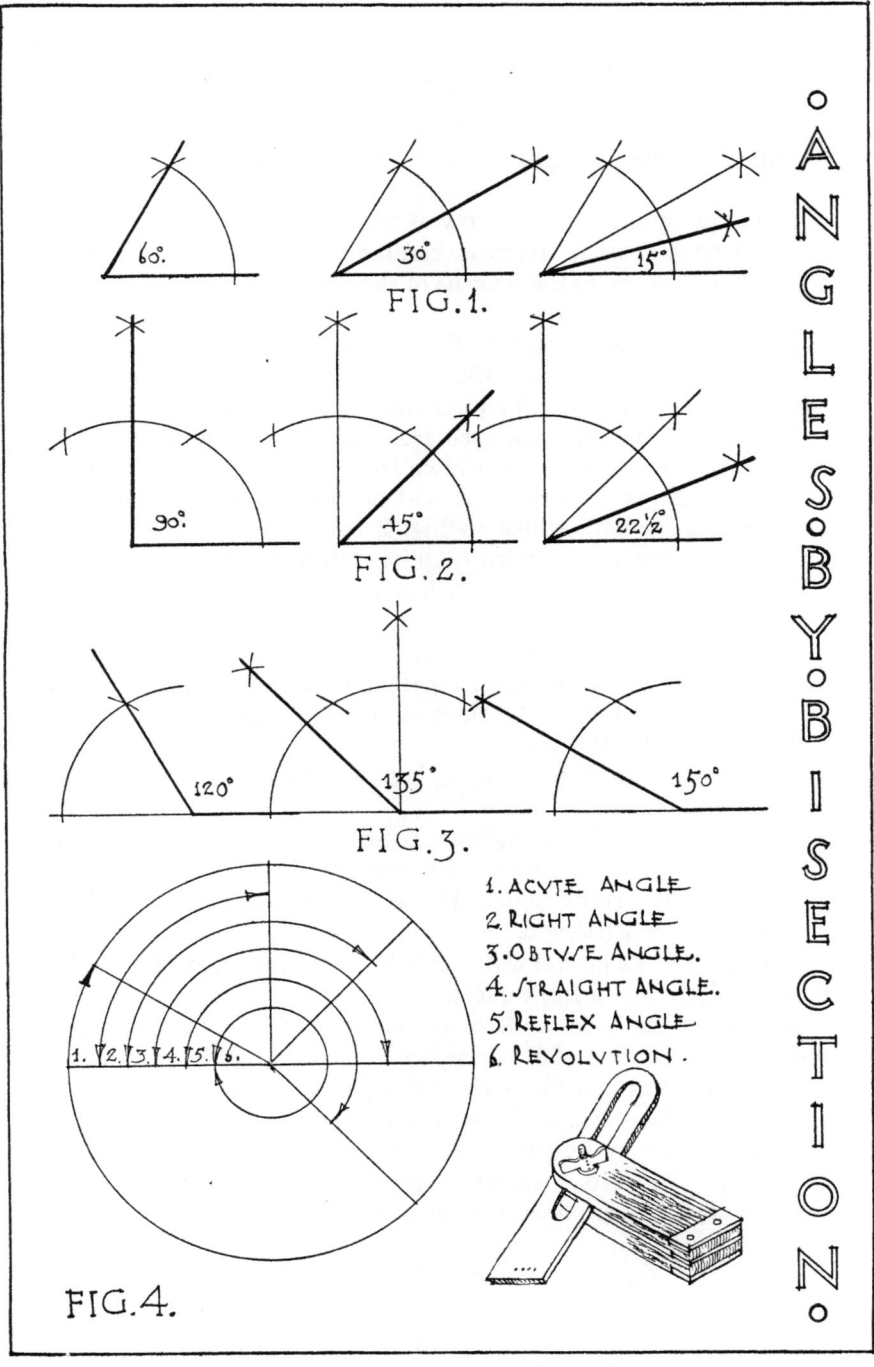

ANGLES AND TRIANGLES

A TRIANGLE, as mentioned under definitions, is a plain figure bounded by three straight lines. Two straight lines cannot enclose a space.

Although an unlimited variety of triangles may be drawn, certain of them have characteristics in common. It is possible to classify triangles according to the lengths of their sides and the sizes of their angles, as follows:

An EQUILATERAL TRIANGLE, Fig. 1, has all its sides equal and all its angles equal. There are 180° in the three angles of any triangle, therefore each angle in an equilateral triangle equals 60°.

The vertical height measured from the base to the vertex is the *altitude* of a triangle. A line drawn from the centre of a side to the opposite angle is a *median*. It will be noticed that in Fig. 1 the altitude is a median, whilst in Fig. 2 where two medians and the altitude are shown they occupy quite different positions.

A SCALENE TRIANGLE, Fig. 2, has all its sides and all its angles unequal.

A RIGHT-ANGLED TRIANGLE, Fig. 3, has one angle that is a right angle. The other two angles together total 90°, but they will be equal only when two of the sides are equal. The side opposite the right angle is the *hypotenuse*.

An ISOSCELES TRIANGLE, Fig. 4, has two of its sides equal and two of its angles equal.

Any two sides of a triangle are together greater than the third.

An OBTUSE-ANGLED TRIANGLE is one in which one of the angles is greater than a right angle. The scalene triangle in Fig. 2 is also an obtuse-angled triangle.

An ACUTE-ANGLED TRIANGLE is one in which each of the three angles is less than a right angle.

Reference has been made to SUPPLEMENTARY ANGLES, two angles which together total 180°; these are illustrated in Fig. 5.

COMPLEMENTARY ANGLES are two angles whose sum equals 90° or one right angle. Each angle is the complement of the other, Fig. 6.

Angle sizes may be transferred without measurement, as shown in Fig. 7. Any arc is struck using the vertex as centre and the length across the arc measured as at OX.

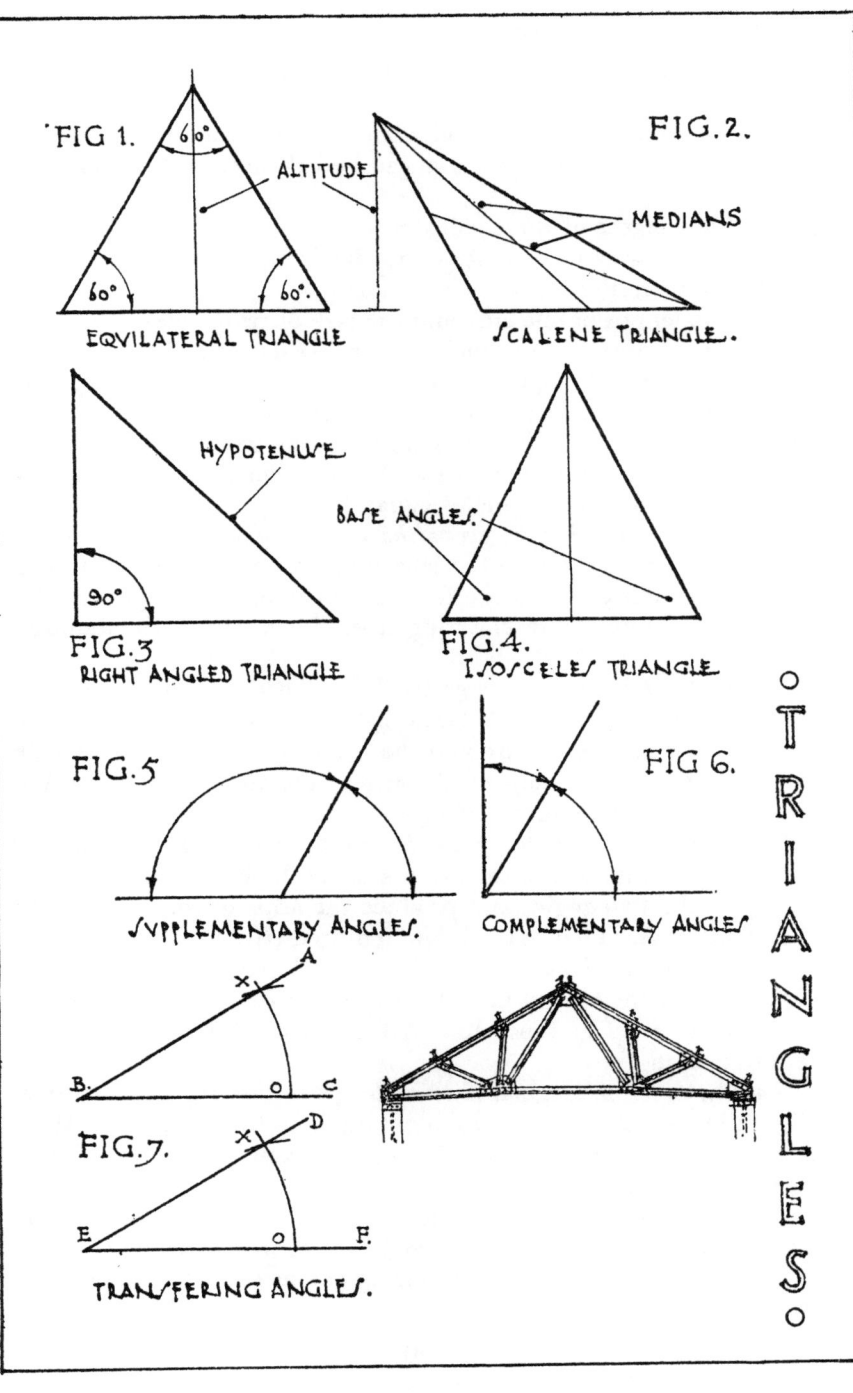

CONSTRUCTION OF TRIANGLES

A TRIANGLE may be constructed from data providing three properties of the triangle are known or can be calculated. They may be either:

(a) The lengths of three sides.

(b) The lengths of two sides and the *included* angle (that made by their intersection).

(c) The length of one side and the size of two angles.

If the type of triangle is known, then less data may be necessary; e.g. an equilateral triangle can be constructed given only the length of one side because we know all the sides are equal and all the angles are 60°. An isosceles triangle may be constructed given one side and one angle, because we know a triangle of this type has two sides equal and two angles equal.

To construct a triangle, given the lengths of three sides, Fig. 1: Draw CB as base equal to the length of one side. With centres C and B and radii equal to the lengths of the other two given sides respectively draw arcs intersecting at A. Then ABC is the required triangle.

To construct a triangle, given the lengths of two sides and the size of the included angle, Fig. 2: Draw the base CB and the given base angle C. Using B as centre and the length of the other given side as radius draw an arc cutting the arm of the angle at A. Join AB. Then ABC is the required triangle.

To construct a triangle, given the length of one side and the size of two angles, Fig. 3: Let the given side be the base of the triangle, construct the two angles and produce the arms of the angles until they meet at A. Then ABC is the required triangle.

To construct an isosceles triangle, given the base and the altitude, Fig. 4: Bisect the base at O and mark the altitude on the perpendicular bisector as at OA. Join AB and AC. Then ABC is the required triangle.

To construct an equilateral triangle, given the length of one side, Fig. 5: Draw the base BC, making it equal to the given side. With centres B and C and radius BC draw arcs intersecting at A. Then ABC is the required triangle.

Knowing the base angles of an equilateral triangle to be 60°, the triangle could have been constructed by erecting these angles at the extremities of the base and producing the arms of the angle until they intersected.

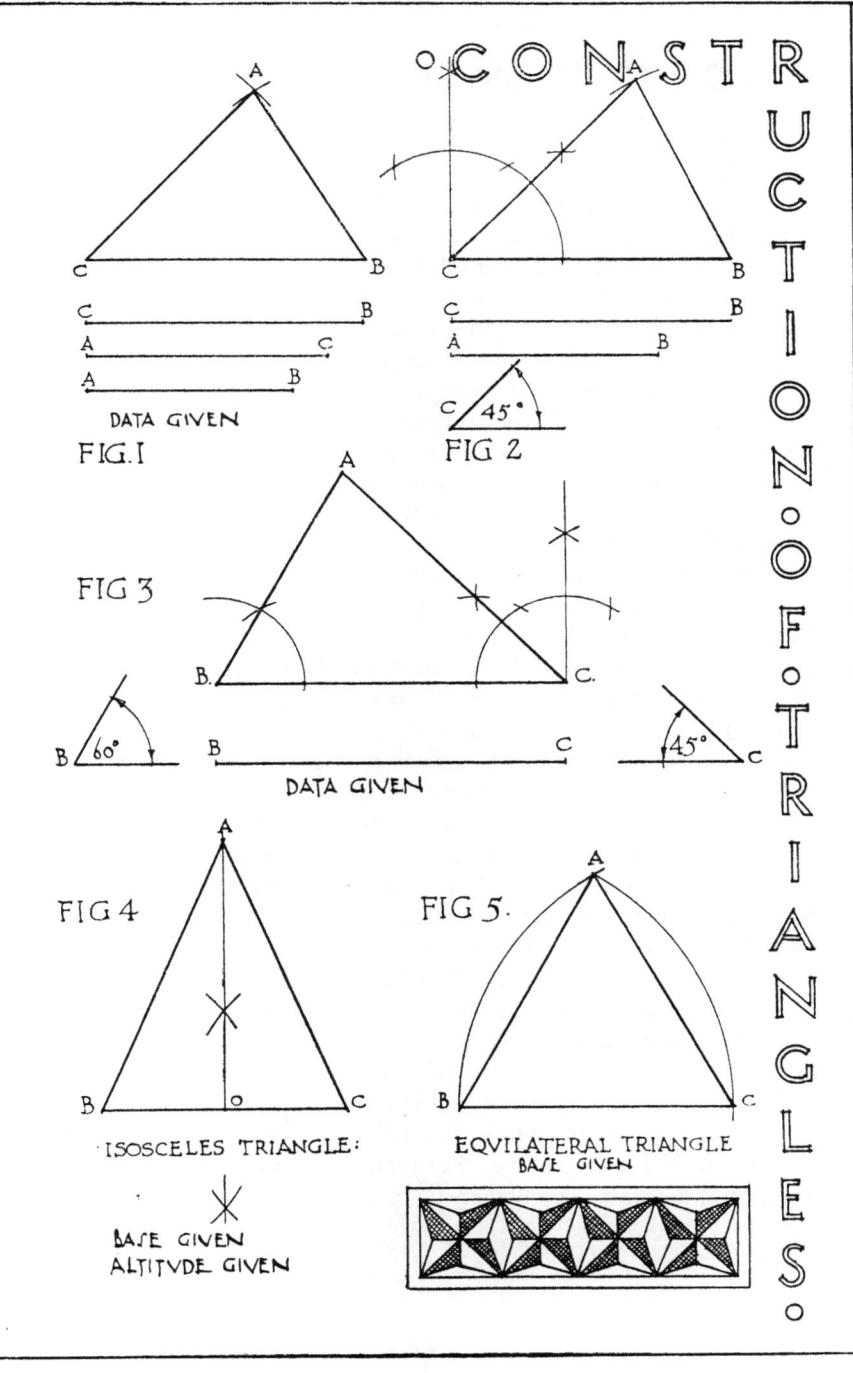

CONSTRUCTION OF TRIANGLES—continued

Some general problems on the construction of triangles are given here:

Fig. 1. *To construct a triangle, given the length of two sides and the altitude of the triangle.* Erect a perpendicular on a base of indefinite length. Mark the length of the altitude on the perpendicular as at AO. With A as centre and the lengths of the two given sides as radii draw arcs cutting the base line in B and C. ABC is the required triangle. (See also exercise on Fig. 4.)

Fig. 2. *To construct a triangle, given the length of the base, one base angle, and the difference between the lengths of the other two sides.* Draw the base BC and the given base angle B. Mark off BD, the difference between the sides, and join DC. The remaining part of the triangle will be isosceles, with base DC and two equal sides. The vertical bisector of DC meets BD produced in A which is the apex of the required triangle ABC.

Fig. 3. *To construct a right-angled triangle, given the length of the hypotenuse and one other side.* Using the hypotenuse as diameter, BC, draw a semicircle. With either B or C as centre and the length of the other given side as radius draw an arc cutting the semicircle at A. Join AB and AC, when ABC is the required right-angled triangle. (The angle in a semicircle is always a right angle; further examples based on this property are shown later.)

Fig. 4. *To draw a triangle, given the lengths of two sides and the angle opposite one of them.* In this example two solutions satisfy the data given. Draw the base line of indefinite length, erect the given angle B and mark off the given line AB. With A as centre and AC, the other given side, as radius strike an arc, which in this case cuts the base in two places C and C[1]. Either of the triangles ABC[1] and ABC satisfy the requirements. Referring back to Fig. 1 it will be seen that there was similarly a second solution to that problem.

Fig. 5. *Any triangle having sides in the proportion 3 : 4 : 5 contains a right angle.* This important fact is often used in the setting out of right angles on the building site, the method being shown in illustration. Building lines and steel tape are used for the sides of the triangle and wood pegs driven in to mark the angle.

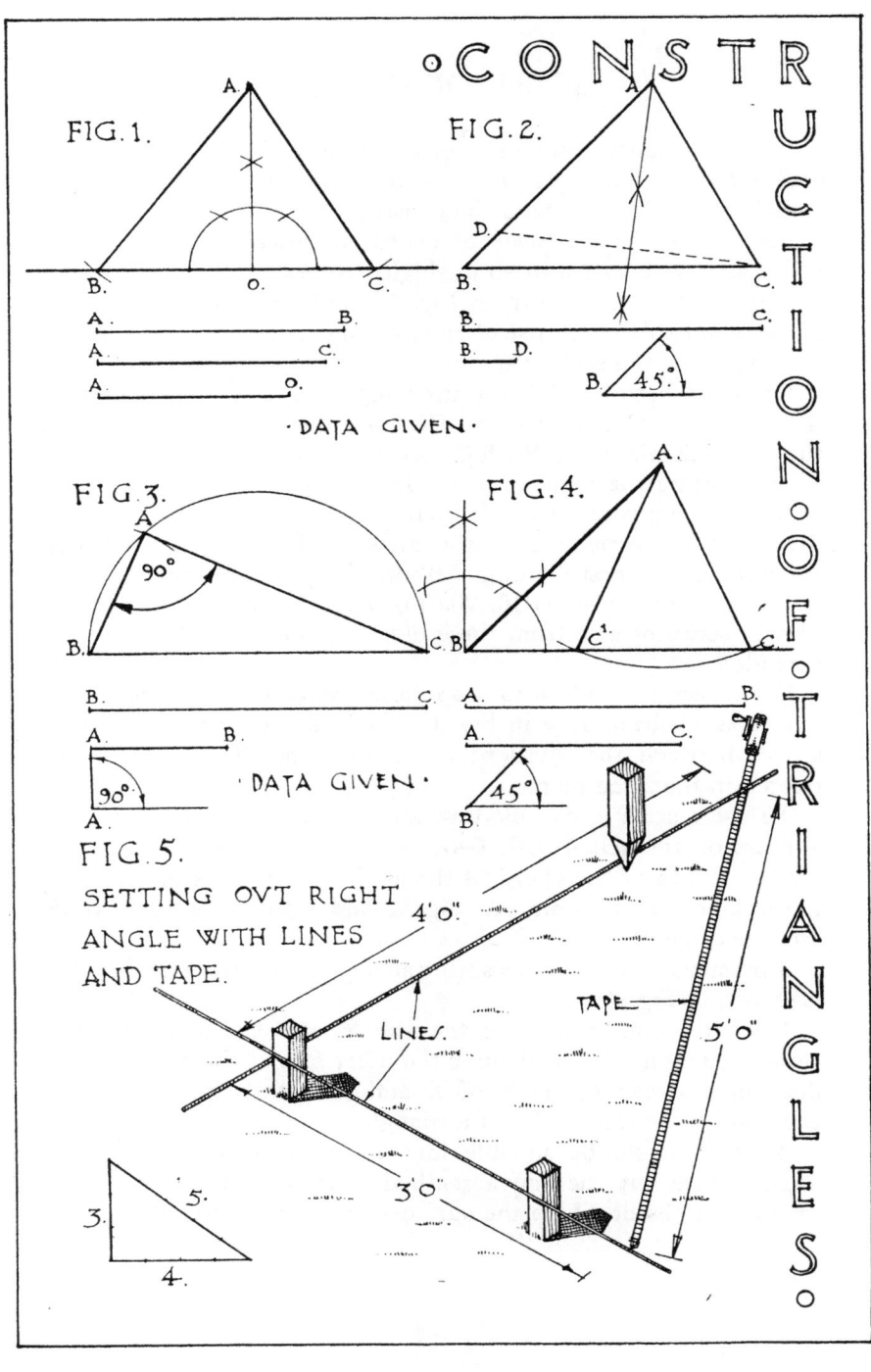

SCALE OF CHORDS

The scale of chords provides a geometrical method of constructing angles that is quite practical for drawing-office work providing sufficient care is taken with the construction.

Many rules have a scale of chords on them, usually marked CHO and graduated 0 to 90 in single degrees.

The construction shown in Fig. 1 enables a similar scale to be drawn and explains the use of any existing scale of chords although the lengths of the scales vary.

Draw the base line CC^1 of any length and at C erect a perpendicular. With C as centre and CC^1 as radius draw an arc cutting the perpendicular at A. With C^1 as centre and the same radius draw an arc cutting the first arc in B. Divide arc AC^1 into nine equal parts, stepping round with dividers. It will be seen that the sixth division falls on point B so the division of the arc can be done more readily by first dividing AB into three equal parts.

Number the divisions 10, 20, 30, etc., and, using C^1 as centre, draw a series of arcs from these divisions onto the base line CC^1 extended.

For clearness it is best to drop these points down vertically onto a scale as has been done in Fig. 1. If odd degrees are required the spaces between the divisions of 10 must be divided again and treated in the same manner.

To use a scale of chords draw an arc with radius CC^1, or as it appears on the actual scale 0–60. Measure from one end of this arc the length of the chord of the angle required, as indicated by the number on the scale. In Figs. 2 and 3 angles of 50° and 35° are constructed.

A practical method of measuring an angle direct from the building is shown in Fig. 4.

A length is marked along the wall as at X and another length along a straight edge held along the other face of the wall, Y. The distance between the ends of X and Y is then measured, thus completing the three sides of a triangle.

A bevel would be suitable for measuring a chamfer or small angular face but the above method gives greater accuracy with greater lengths or where the surfaces are slightly irregular.

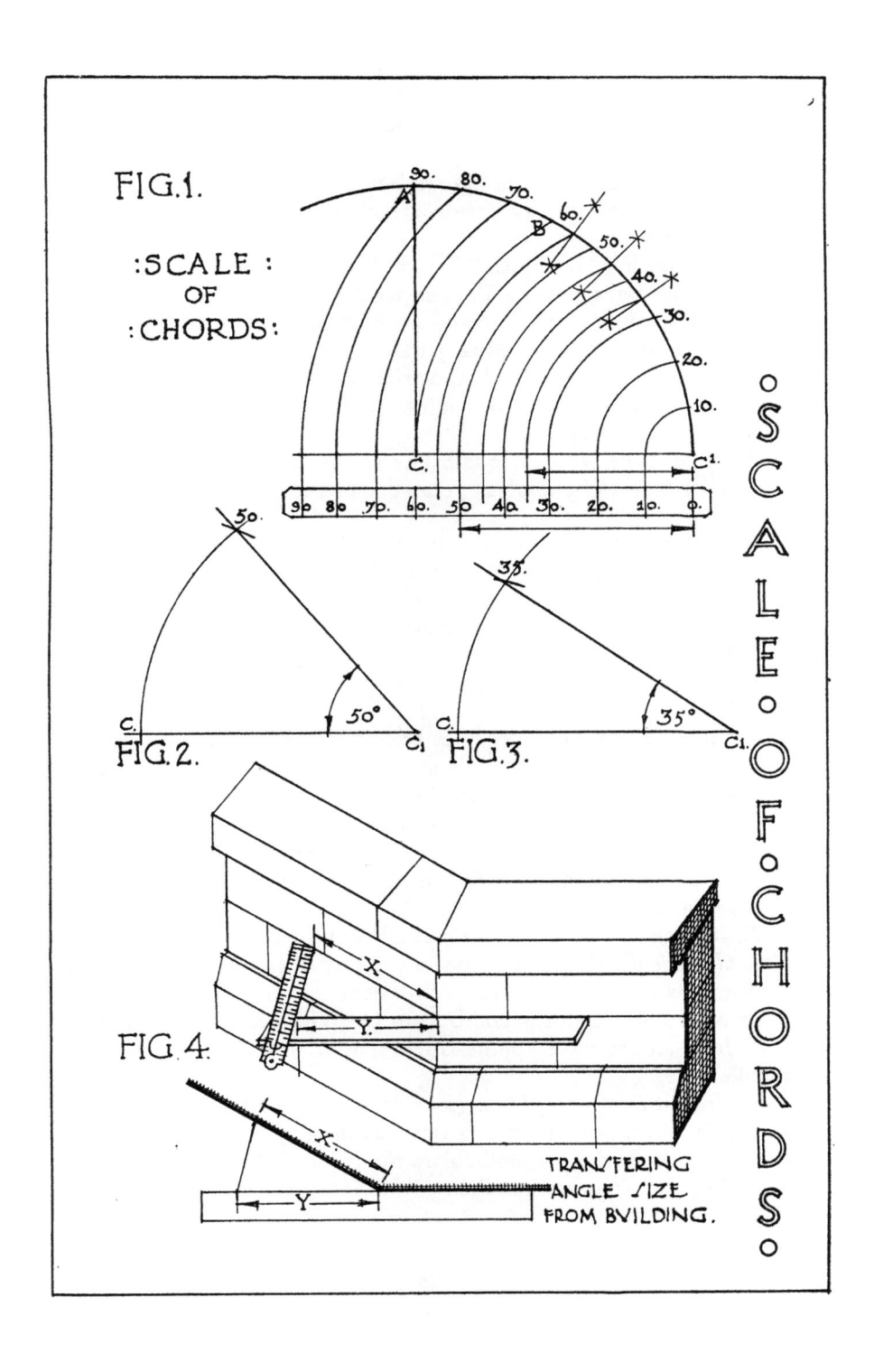

FIG. 1.

: SCALE :
OF
: CHORDS :

FIG. 2.   50°

FIG. 3.   35°

FIG. 4.

TRANSFERING
ANGLE SIZE
FROM BUILDING.

SCALE ○ OF ○ CHORDS

QUADRILATERALS AND PARALLEL LINES

QUADRILATERALS are plane figures bounded by four straight lines. The four angles in any quadrilateral together equal four right angles or 360°.

A SQUARE, Fig. 1. A quadrilateral having all its sides equal and all its angles right angles. The *diagonals* (straight lines drawn from corner to corner) of a square bisect each other at right angles and divide the area into four equal parts.

A RECTANGLE, Fig. 2. A quadrilateral with its opposite sides equal and all its angles right angles. Its diagonals bisect each other but not at right angles.

A RHOMBUS, Fig. 3. A quadrilateral with all its sides equal but its angles not right angles. The diagonally opposite angles are equal.

A PARALLELOGRAM, Fig. 4. Any quadrilateral having its opposite pairs of sides parallel. (Note that the square and rhombus are particular kinds of parallelogram.)

A TRAPEZIUM, Fig. 5. A quadrilateral having two only of its sides parallel.

A TRAPEZOID, Fig. 6. A quadrilateral having no two of its sides parallel.

PARALLEL LINES. Parallel lines are straight lines which lie in the same plane and are the same distance apart throughout their length. If produced in either direction they will never meet.

To draw a line parallel to a given line AB *through a given point* X, Fig. 7: Draw a line from X to any point C on AB. Measure angle XCB and construct angle DXC equal to it. Produce DX when DE will be parallel to AB.

To draw a line parallel to a given line AB *from a given point* X. *Second method*, Fig. 8: Choose any point C in the line AB and with radius CX draw arc XD. With centre X and the same radius draw an arc of indefinite length from C. With centre C and radius DX draw an arc fixing point E. Join EX, which is the required parallel line.

To draw a line parallel to a given line AB *at a given distance from it*, Fig. 9: Erect perpendiculars towards the ends of AB. By means of arcs mark on the perpendiculars the given distances as at C and D. CD is the required parallel.

FIG.1.
: SQUARE :

FIG.2.
: RECTANGLE :

DIAGONALS.

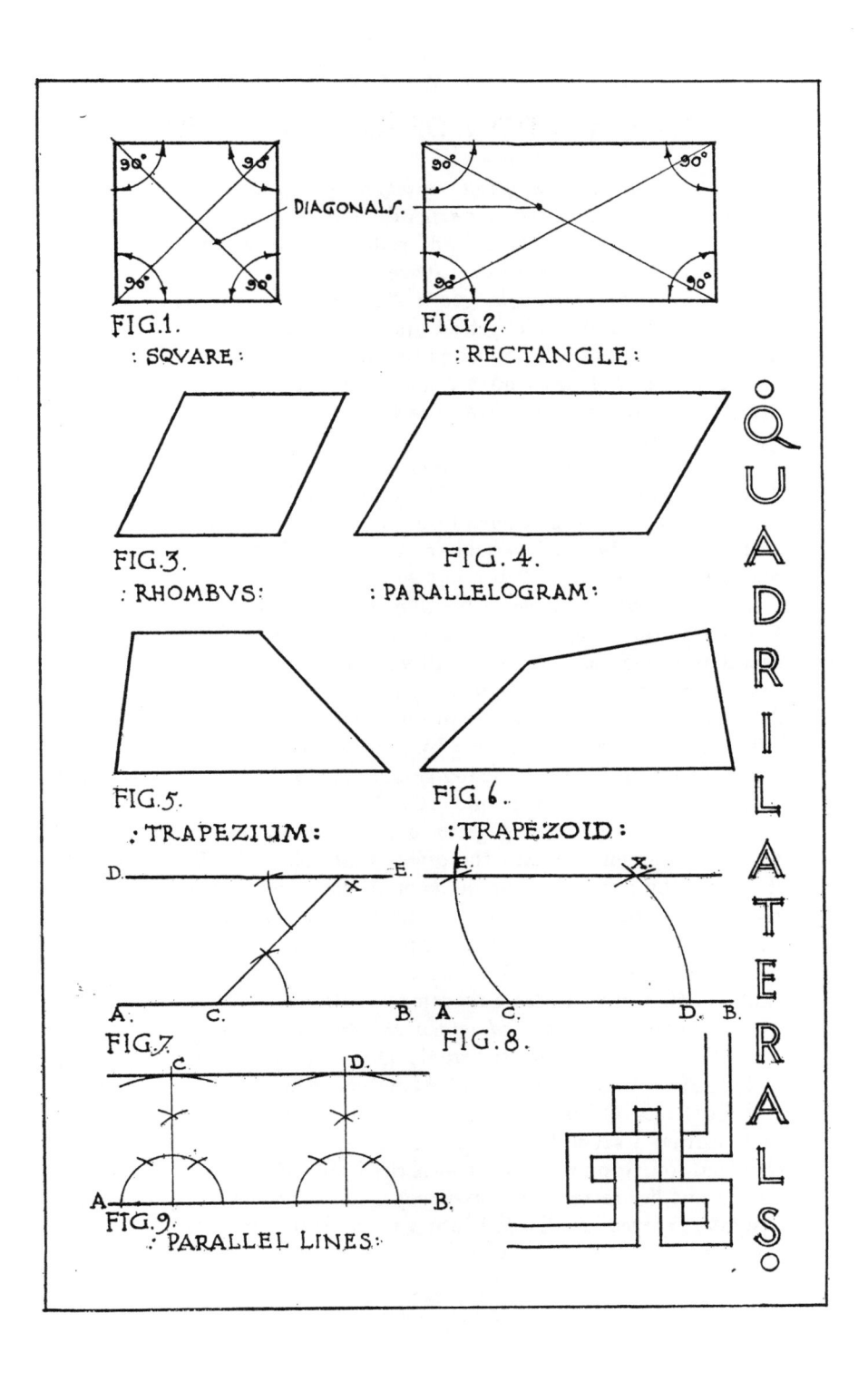

FIG.3.
: RHOMBUS :

FIG.4.
: PARALLELOGRAM :

FIG.5.
: TRAPEZIUM :

FIG.6.
: TRAPEZOID :

FIG.7.

FIG.8.

FIG.9.
: PARALLEL LINES :

QUADRILATERALS.

CONSTRUCTION OF QUADRILATERALS

To construct a square, given the length of one side: Let AB be the given side, Fig. 1. Erect a perpendicular at A, making CA equal to AB. With centres B and C and radius AB draw arcs intersecting at D. ABCD is the required square.

To construct a rectangle, given the length of two sides, Fig. 2: Draw AB equal to the length of one given side. Erect the perpendicular CA, making CA equal to the other given side. With centre C and radius AB draw an arc and with centre B and radius AC draw a second arc intersecting the first arc at D. ABCD is the required rectangle.

To construct a square on a given diagonal: Let AB be the given diagonal, Fig. 3. Bisect AB with a perpendicular at the point O. As the diagonals of a square bisect each other at right angles CO and DO must be made equal to AO and BO, then ABCD is the required square.

To construct a rectangle, the diagonal and one side being given: Let AB be the given diagonal and AC the length of the given side. Bisect the diagonal AB in O and with O as centre and radius AO draw a circle. As the angle in a semicircle is a right angle, the diagonal divides the rectangle into two equal right-angled triangles. These triangles may be drawn by cutting the circumference of the circle by arcs at C and D, using centres A and B and a radius equal to the length of the given side AC. (Fig. 4).

To construct a rhombus, given one side and one base angle: Let AB be the given side and the given angle 60°, Fig. 5. After constructing the given base angle as at A the same methods are used as in Figs. 1 or 2, remembering that a rhombus has all its sides of equal length and its diagonally opposite angles equal. (Angle A + Angle D therefore = 120°; Angle C + Angle B must be 360° − 120° = 240°; each of these therefore = 120°.)

To construct a trapezoid, given the length of four sides and the size of one angle: Draw the angle CAB equal to the given angle making AB equal to side 1 and AC equal to side 2. With centre C and side 3 as radius draw an arc. With centre B and side 4 as radius draw a second arc intersecting the first in D, which is the required fourth point. (NOTE.—If the order of the sides is changed a different figure results. In surveys, to overcome this possibility, irregular figures are divided into a number of triangles.)

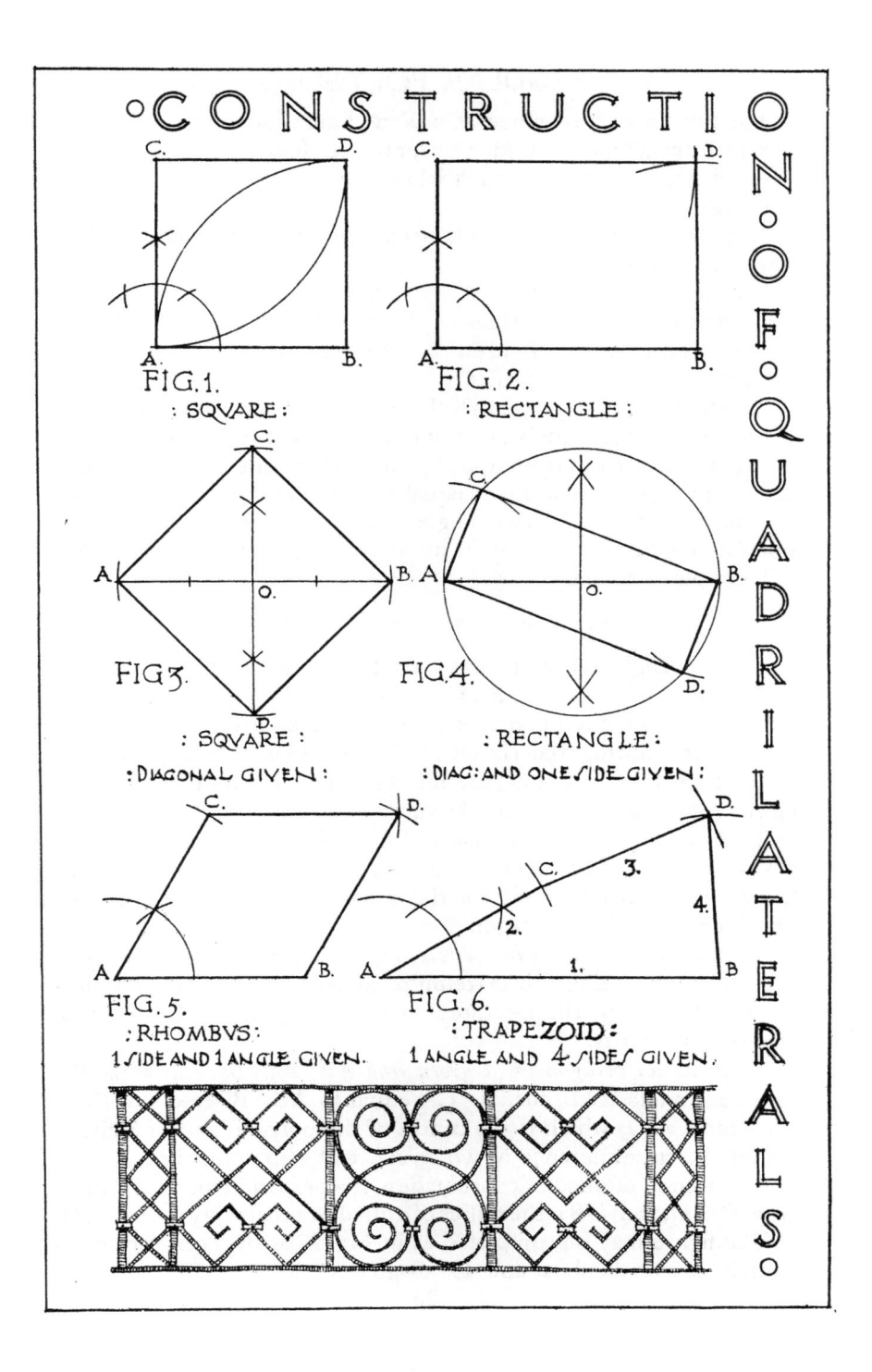

CONSTRUCTION OF QUADRILATERALS

FIG. 1.
: SQUARE :

FIG. 2.
: RECTANGLE :

FIG. 3.
: SQUARE :
: DIAGONAL GIVEN :

FIG. 4.
: RECTANGLE :
: DIAG: AND ONE SIDE GIVEN :

FIG. 5.
: RHOMBUS :
1 SIDE AND 1 ANGLE GIVEN.

FIG. 6.
: TRAPEZOID :
1 ANGLE AND 4 SIDES GIVEN.

REGULAR POLYGONS

A POLYGON may be defined as a plane figure having many angles, or a rectilineal figure bounded by more than four sides. An *irregular polygon* has its sides unequal, whilst a *regular polygon* has all its sides and angles equal.

Regular polygons are named according to the number of their angles and sides.

PENTAGON	. .	5 sides	NONAGON	. .	9 sides
HEXAGON	. .	6 sides	DECAGON	. .	10 sides
HEPTAGON	. .	7 sides	UNDECAGON	.	11 sides
OCTAGON	. .	8 sides	DUODECAGON	.	12 sides

The number of angles equals the number of sides, and the sum of the external angles equals the sum of the internal angles, 360°, Fig. 3.

Lines drawn from the angular points to the centre of the polygon divide the figure into as many equal triangles as the figure has sides.

The size of the internal angle between any two sides can be obtained by taking the size of the internal angle at the centre from 180°.

Example, octagon:

$$\text{Internal angle at centre} = \frac{360°}{8} = 45°$$

$$\text{Angle between sides} = 180° - 45° = 135°$$

To draw a hexagon in a circle, Fig. 1: The radius of a circle steps round the circumference six times which gives the necessary points for drawing the six-sided figure. This fact must not be confused with π, 3·14159, the number of times the diameter of a circle divides into the circumference.

To draw an octagon in a square, Fig. 2: Using the corners of the square ABC and D as centres and half the diagonal (AO) as radius, draw arcs cutting the sides of the square at points which are the angular points of the octagon.

To draw a duodecagon in a circle, Fig. 4: Draw the two diameters AB and CD. Using the extremities as centres and AO as radius draw arcs cutting the circumference which will be divided into twelve equal parts.

To draw an octagon on a given line AB, Fig. 5: Construct the first 135° angle CAB, making CA equal to AB. Bisectors of these lines meet at O, which is the centre of the polygon and of a circle, round the circumference of which AB will step eight times.

As a second method of construction, if a perpendicular is dropped from C to D, then DB is the radius of a quadrant that will give centre O.

The figure may be completed as in Fig. 2, or by a series of horizontals, perpendiculars, and 45° diagonals.

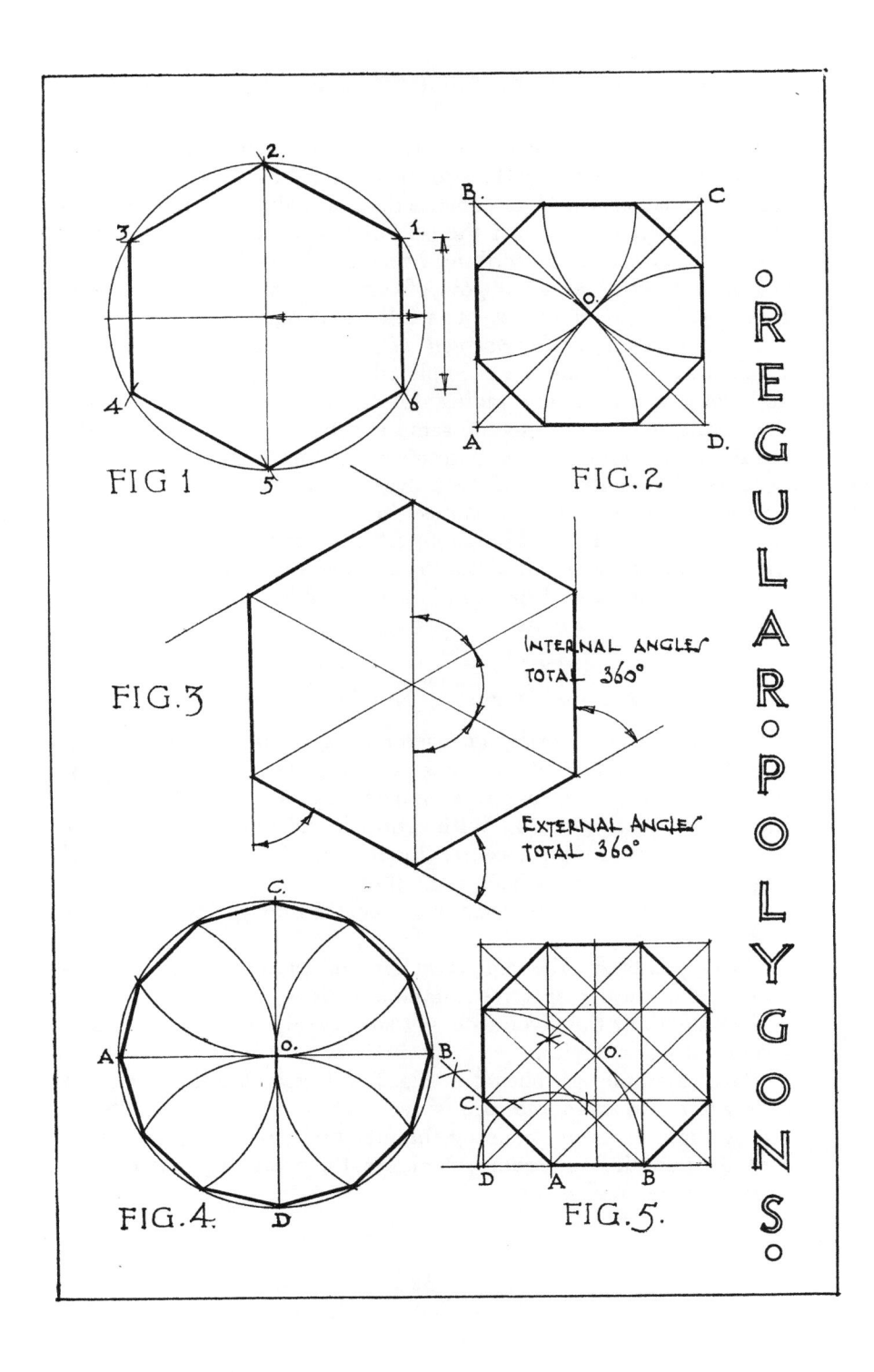

REGULAR POLYGONS

FIG 1

FIG. 2

FIG. 3

INTERNAL ANGLES
TOTAL 360°

EXTERNAL ANGLES
TOTAL 360°

FIG. 4.

FIG. 5.

CONSTRUCTION OF REGULAR POLYGONS

THE two examples shown here are suitable constructions for any regular polygon whereas the previous examples were suitable only for particular types. The constructions are also suitable for portions of polygons providing the work is not on so large a scale that the centre of the circumscribing circle is inaccessible.

For portions of large polygons of this type, such as bay windows, it is usually more convenient to work from angles.

To construct a regular polygon on a given side: Let AB be the given side, Fig. 1, and the example taken a heptagon. Produce AB and with centre A and radius AB draw a semicircle. By trial divide the semicircle into the same number of equal parts as the required polygon has sides, 7 in this case. Join A to point 2, which becomes the second side of the polygon.

Bisectors of these two sides meet in O which is the centre of the circumscribing circle. The length AB may now be stepped round the circumference of this circle to complete the figure.

NOTE.—The second division is always taken because the external angle of any polygon

$$= \frac{360°}{\text{Number of sides}} \text{ or } \frac{180°}{\text{Number of sides}} \times 2.$$

To construct any number of regular polygons on a given side AB, Fig. 2: Bisect AB which becomes the centre line. Erect a perpendicular BC from B, making its length equal to AB. Join AC, cutting the centre line in point 4. With centre B and radius AB draw the quadrant AC, cutting the centre line in point 6. Bisect the vertical height between points 4 and 6, giving point 5.

Step the distance 4 to 5 along the centre line, giving the points 7, 8, 9, 10, etc.

These points are the centres of circumscribing circles of polygons having these respective numbers of sides.

The length of the given side will step evenly round the circumference of these circles.

Three examples are shown in Fig. 2: a hexagon, a nonagon, and a decagon. Considerable care is necessary for accurate results, as a very slight error in obtaining the first division is magnified as it ascends the scale. In practice, therefore, the preceding construction is often more convenient.

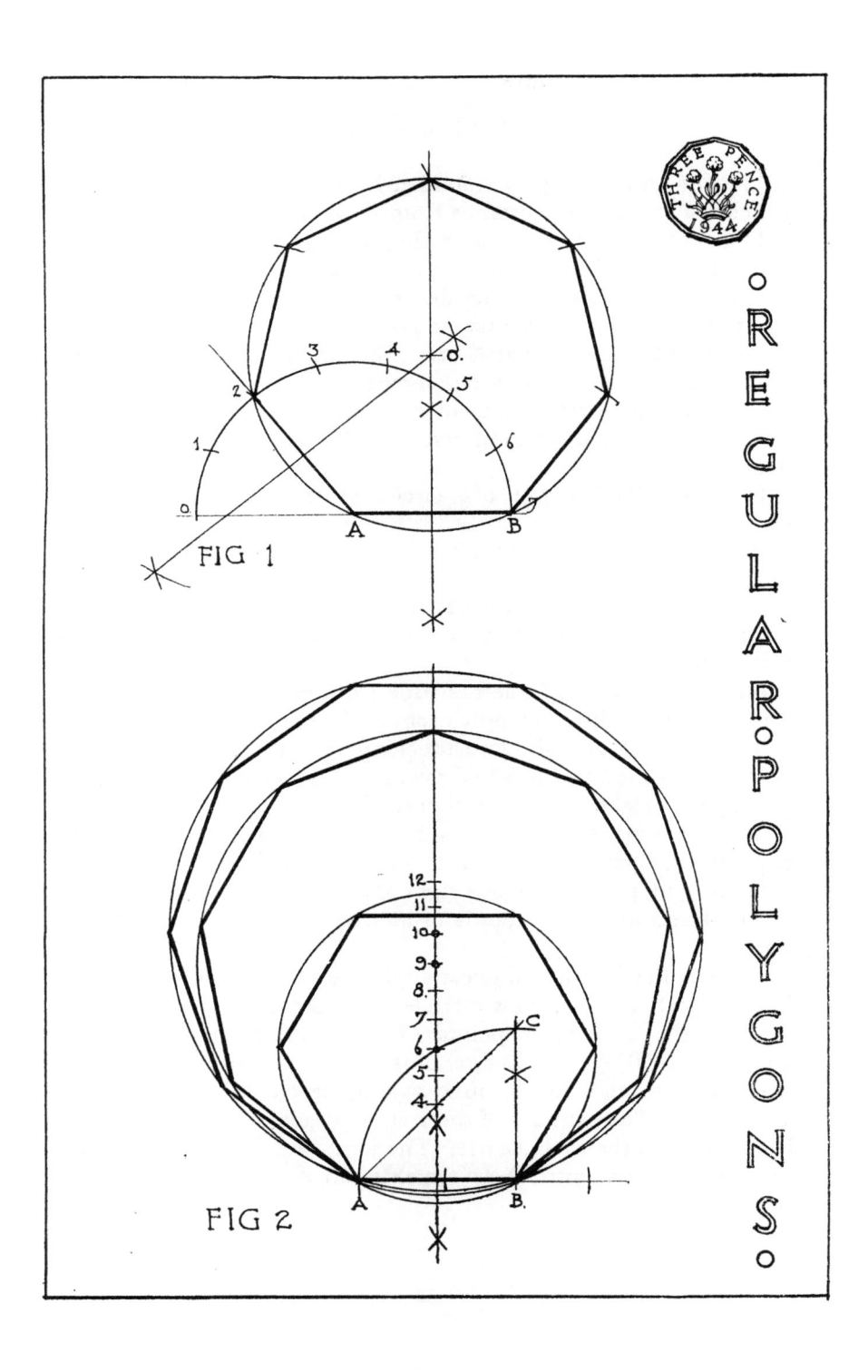

FIG 1

FIG 2

REGULAR POLYGONS

THE CIRCLE

A CIRCLE is a plane figure bounded by a curved line all points in which are at the same distance from the centre.

DIAMETER. A straight line passing through the centre of a circle and cutting the circumference at both ends.

RADIUS (plural *radii*). The distance from the centre to the circumference. It is half the diameter.

CIRCUMFERENCE (or PERIMETER). The line bounding the circle.

ARC. A portion of the circumference.

CHORD. Any straight line in a circle, both ends of which are terminated by the circumference. The greatest chord is the diameter.

SEGMENT. The portion of a circle enclosed by a chord and the corresponding arc.

SEMICIRCLE. One of the two halves into which a circle is cut by its diameter.

SECTOR. A portion of a circle enclosed by two radii and an arc.

QUADRANT. A sector bounded by two radii at right angles; it is a quarter of the circle.

NORMAL. A straight line radiating from the centre, i.e. the continuation of a radius beyond the circumference.

TANGENT. A straight line touching the circumference at one point only called the *point of contact*. A tangent at the point of contact is at right angles to the normal.

The angle in a segment is constant. In Fig. 3 angle ACB in the segment on chord AB is equal to any other such angle, e.g. ADB. The angle in a semicircle is a right angle, Fig. 4.

The bisectors of the chords of a circle pass through its centre, Fig. 5.

Fig. 6 is the practical application of this rule to a segmental arch in brickwork, showing how a circle may be drawn which will pass through any three given points. The span AB and rise CD are given. AB, CB, AC are each chords of a circle and their bisectors meet in O, which is the required striking centre. The radius of the arch is OC. The extrados of the arch is parallel to the intrados and is struck from the same centre. The joints are normals and therefore radiate from centre O, as joints should always do for maximum strength.

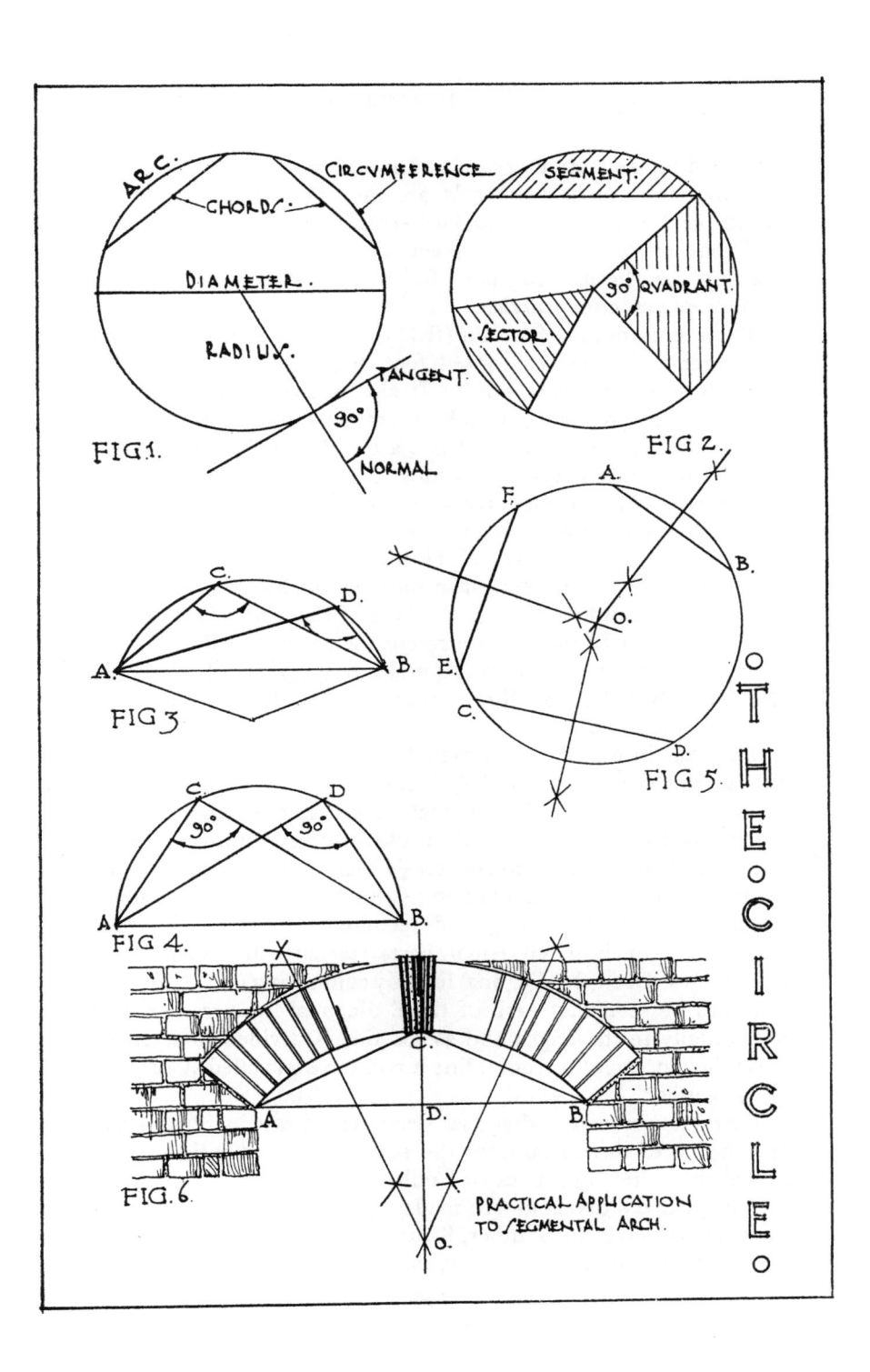

FIG 1.
ARC. CIRCUMFERENCE
CHORDS.
DIAMETER.
RADIUS.
TANGENT.
90°
NORMAL

FIG 2.
SEGMENT.
90° QUADRANT.
SECTOR.

FIG 3.
A. C. D. B. E.

FIG 5.
A. F. B. O. E. C. D.

FIG 4.
C. 90° 90° D.
A. B.

FIG. 6.
A. C. D. B. O.
PRACTICAL APPLICATION TO SEGMENTAL ARCH.

THE CIRCLE

THE CONSTRUCTION OF SCALES

In building and architectural work scales ranging from $\frac{1}{16}$ inch = 1 foot to 3 inches = 1 foot are in general use and are to be found on the architect's standard scale rule. On a scale of 1 inch = 1 foot, every foot on the object is represented by 1 inch on the drawing, i.e. the drawing is $\frac{1}{12}$ full size; this fraction $\frac{1}{12}$ is known as the *representative fraction*.

The representative fraction (R.F.) is usually mentioned only on the larger scales, $1\frac{1}{2}$ inches = 1 foot, being $\frac{1}{8}$ full size ($\frac{1}{8}$ F.S.) or 3 inches = 1 foot being $\frac{1}{4}$ F.S. It also is used in small scales for survey plans and maps. 1 inch to a mile would be represented by the fraction 1/63360 because there are 63360 inches in one mile.

It is sometimes necessary to construct a scale other than those in general use. The problem resolves itself into the geometrical problem of dividing a line into any given number of equal parts.

In Figs. 1 and 2 scales of $1\frac{1}{2}$ inches and $2\frac{1}{4}$ inches to 1 foot are constructed. First set out a number of divisions $1\frac{1}{2}$ inches in length representing feet (Fig. 1). One of these divisions must now be divided into 12 parts to represent inches as at AB.

Draw CA at any angle to AB and any length easily divisible by 12. Join CB and make all the other lines parallel to it, cutting AB in equal divisions.

The scale in Fig. 2 is constructed in a similar manner after setting out the primary divisions of $2\frac{1}{4}$ inches.

DIAGONAL SCALES. The diagonal scale enables smaller sizes to be accurately drawn and is particularly useful when working in decimals.

In Fig. 3 a scale to read inches, $\frac{1}{10}$ inch and $\frac{1}{100}$ inch is shown.

Divide the scale into inches and $\frac{1}{10}$ inch by the previous method, these being called primary and secondary divisions. Divide the vertical height into ten equal parts (tertiary divisions) by horizontal lines. Draw the diagonal lines by connecting 0–1, 1–2, 2–3, etc.

It will be seen that each of these diagonal lines travels $\frac{1}{10}$ inch horizontally in its length and as the line is divided into ten equal parts then at each horizontal line it will have travelled $\frac{1}{10}$ of $\frac{1}{10}$ inch, i.e. $\frac{1}{100}$ inch.

Further uses of the diagonal scale are shown in Figs. 4 and 5. In the first of these examples the scale is to read miles, furlongs, and chains when $1\frac{1}{2}$ inches = 1 mile.

The primary divisions are made $1\frac{1}{2}$ inches long, the secondary divisions are eight in number, 8 furlongs equalling 1 mile, and the tertiary divisions ten in number, there being 10 chains in 1 furlong.

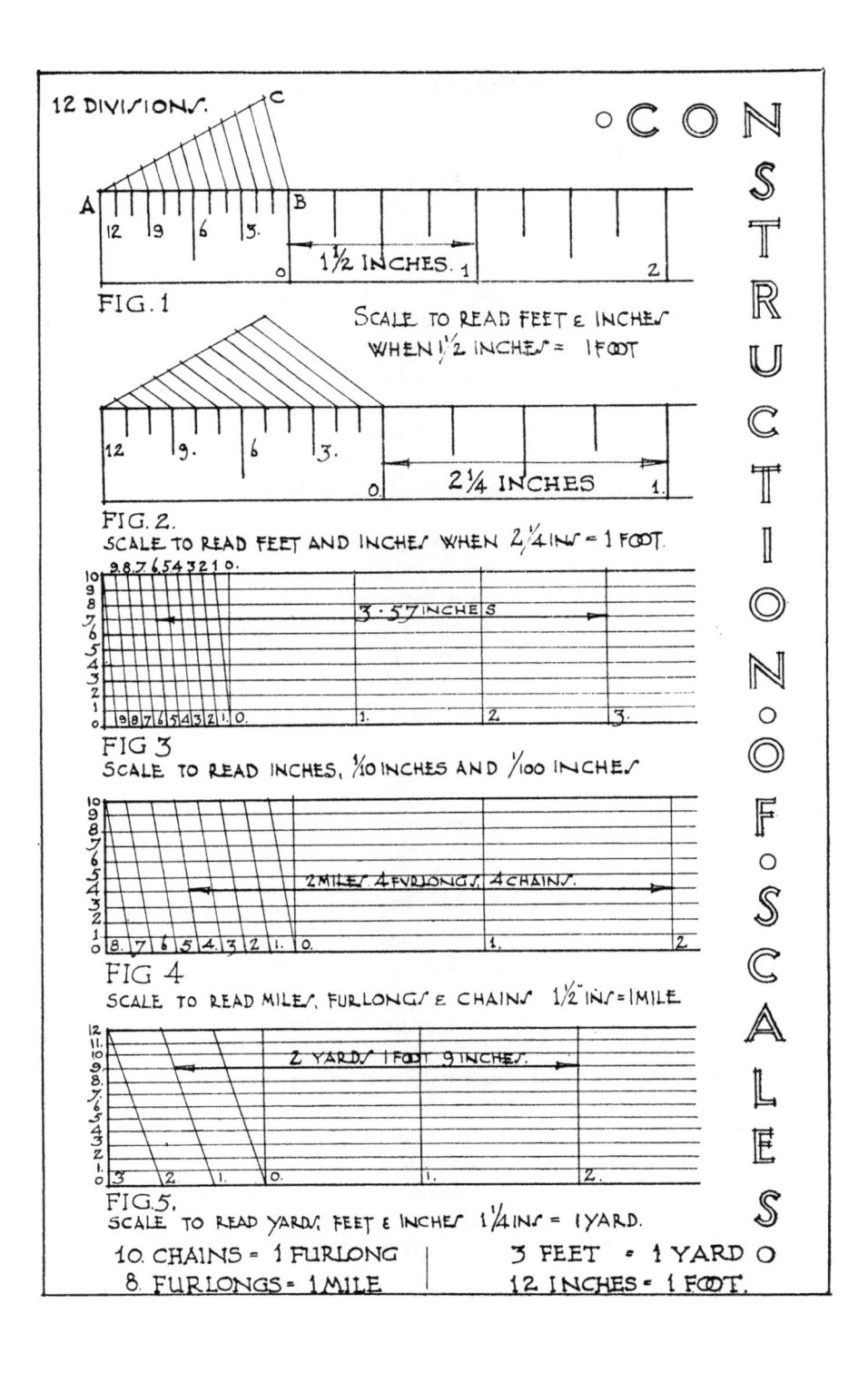

CONSTRUCTION OF SCALES

FIG. 1
12 DIVISIONS.
1½ INCHES.

FIG. 2.
SCALE TO READ FEET AND INCHES WHEN 2¼ INS = 1 FOOT.
SCALE TO READ FEET & INCHES WHEN 1½ INCHES = 1 FOOT
2¼ INCHES

FIG 3
SCALE TO READ INCHES, 1/10 INCHES AND 1/100 INCHES
3·57 INCHES

FIG 4
SCALE TO READ MILES, FURLONGS & CHAINS 1½ INS = 1 MILE
2 MILES 4 FURLONGS 4 CHAINS.

FIG. 5.
SCALE TO READ YARDS, FEET & INCHES 1¼ INS = 1 YARD.
2 YARDS 1 FOOT 9 INCHES.

10 CHAINS = 1 FURLONG | 3 FEET = 1 YARD
8 FURLONGS = 1 MILE | 12 INCHES = 1 FOOT.

ENLARGEMENT BY SQUARES

IRREGULAR figures, particularly maps, may usually be enlarged or reduced proportionally by the simple process of dividing the original into a number of squares of equal size and transferring points onto a new series of squares enlarged or reduced according to the scale required.

In Fig. 2 a portion of the map of England is represented, the height of the original being 3 inches.

It is required to draw a new map having a height of 4 inches.

Divide the 3-inch side into a number of equal parts (twelve taken) and by means of vertical divisions of the same size construct a sufficient number of squares to enclose the map.

Draw the vertical side of the new map 4 inches and divide this also into twelve equal parts, completing the squares by divisions of a similar size.

A fair approximation of the irregular curve can be obtained by transferring intersecting points from one series of squares to the other.

Point Z is indicated on both maps and is at the intersection of vertical line 9 and horizontal line 1. A point Y is also shown which is not actually at any intersection. In this case the distance of the point below horizontal would be judged, or if greater accuracy is required then a greater number of original squares would have to be taken.

It will be found that graph paper is particularly useful for this purpose.

PROPORTIONAL SCALE. The proportional scale enables lengths to be enlarged or reduced proportionally without resorting to calculations.

The setting-out of the scale in Fig. 1 has been adapted to suit the map in Figs. 2 and 3 and the distances X and X_1, A and A_1 are proportional both on the scale and by the previously shown method of squares.

It is required to construct a scale to enlarge or reduce to the proportion 4 : 3.

Draw a base line and a vertical on it at any convenient point S. Mark off on either side of S equal divisions to the number of four and three respectively. In the example these have been further subdivided, giving lines SQ and SR of eight divisions and six divisions respectively. Taking any point P on the vertical, draw the triangle PQR. Any horizontal line drawn between PQ and PR will be divided in the required proportion by the vertical.

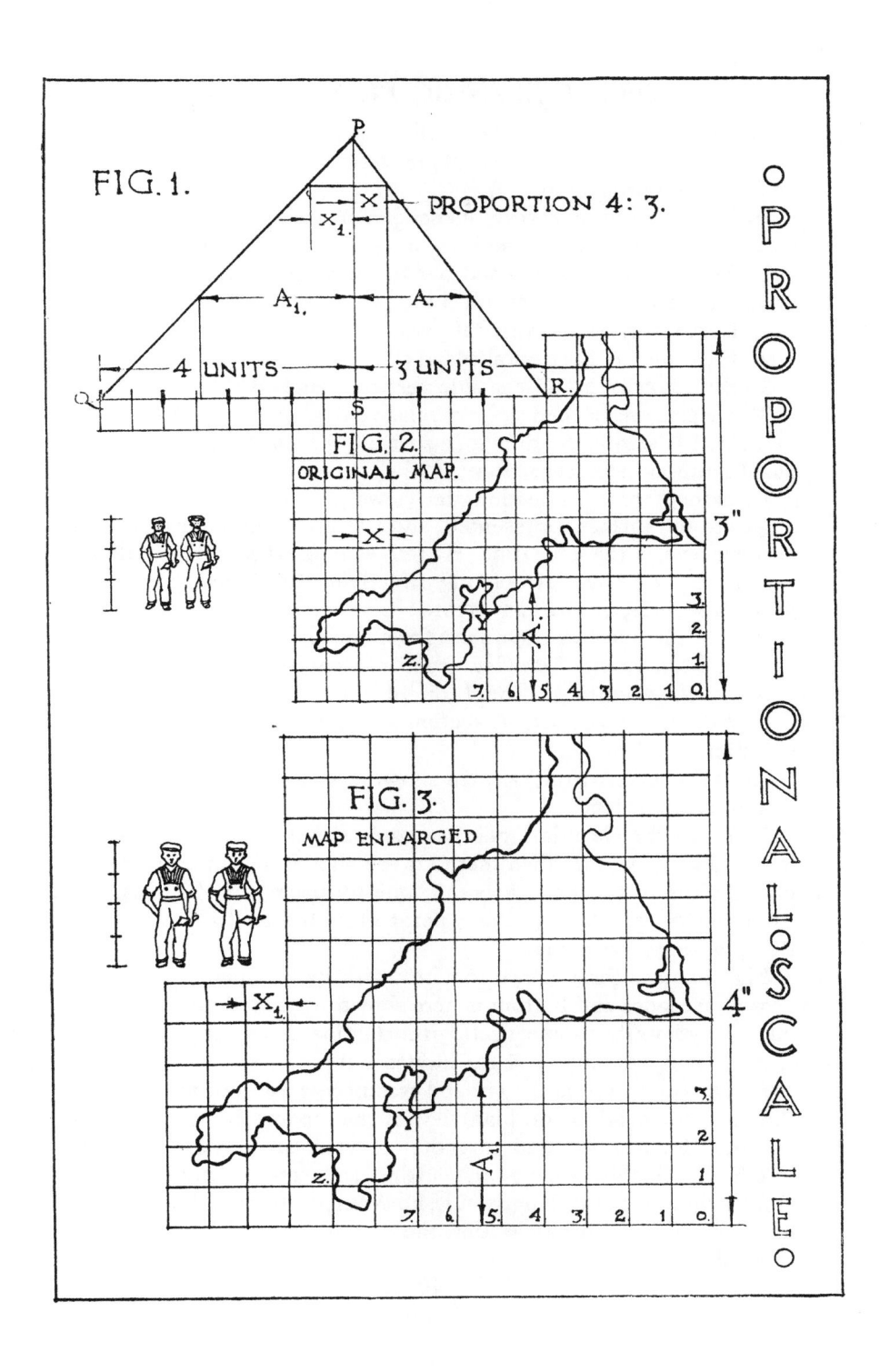

FIG. 1.

PROPORTION 4 : 3.

4 UNITS   3 UNITS

FIG. 2.
ORIGINAL MAP.

3"

FIG. 3.
MAP ENLARGED

4"

PROPORTIONAL SCALE

ORTHOGRAPHIC PROJECTION

THE method of representing a solid object on a plane surface by means of independent views of its various surfaces is known as orthographic projection.

In Fig. 3 a box is shown suspended in space with views of its surfaces projected on to the imaginary planes of projection.

If these planes were opened out to form one surface, as a sheet of paper, then the views would be:
1. ELEVATION on the vertical plane (V.P.).
2. PLAN on the horizontal plane (H.P.).
3. SIDE ELEVATION on the side vertical plane (S.V.P.).

The views are in their correct relative positions and an object so presented is said to be drawn in *first angle projection*, which is the usual method of presentation.

In orthographic projection the views are always taken at right angles to the surface represented, and it is assumed that every point is at eye level at the same time. We may examine these views in detail:

ELEVATION. Any view looking directly at the object; it may be a *front, back, side,* or *end* ELEVATION.

PLAN. The view of the object from above.

REFLECTED PLAN. The view from underneath.

SECTION. A cutting through the object.

SECTIONAL ELEVATION. A section which also includes part of an elevation.

HORIZONTAL SECTION. A plan showing a section made by a horizontal cutting plane.

In Fig. 1 the elevation, plan, and side elevation of two intersecting girders is given. Two methods are shown for transferring the plan widths to the side elevation: one by means of 45° lines, and the other by rebatment, using point O as the centre from which the quadrants are struck.

Fig. 2 shows three views of a box with its lid open. The thickness of the sides and bottom is here shown by dotted lines, but may be omitted unless some useful information is thereby conveyed. Dotted lines are only used for hidden outlines.

A vertical section is shown, and was necessary in order to obtain the correct elevation and plan view of the top and bottom edges of the lid. Its position and direction of view are indicated by the line X–X. In this case it is one cutting, the section of the box being similar throughout its length; it is, however, quite usual for economy of effort to show in one section the details revealed by cuttings at more than one place.

ORTHOGRAPHIC PROJECTION

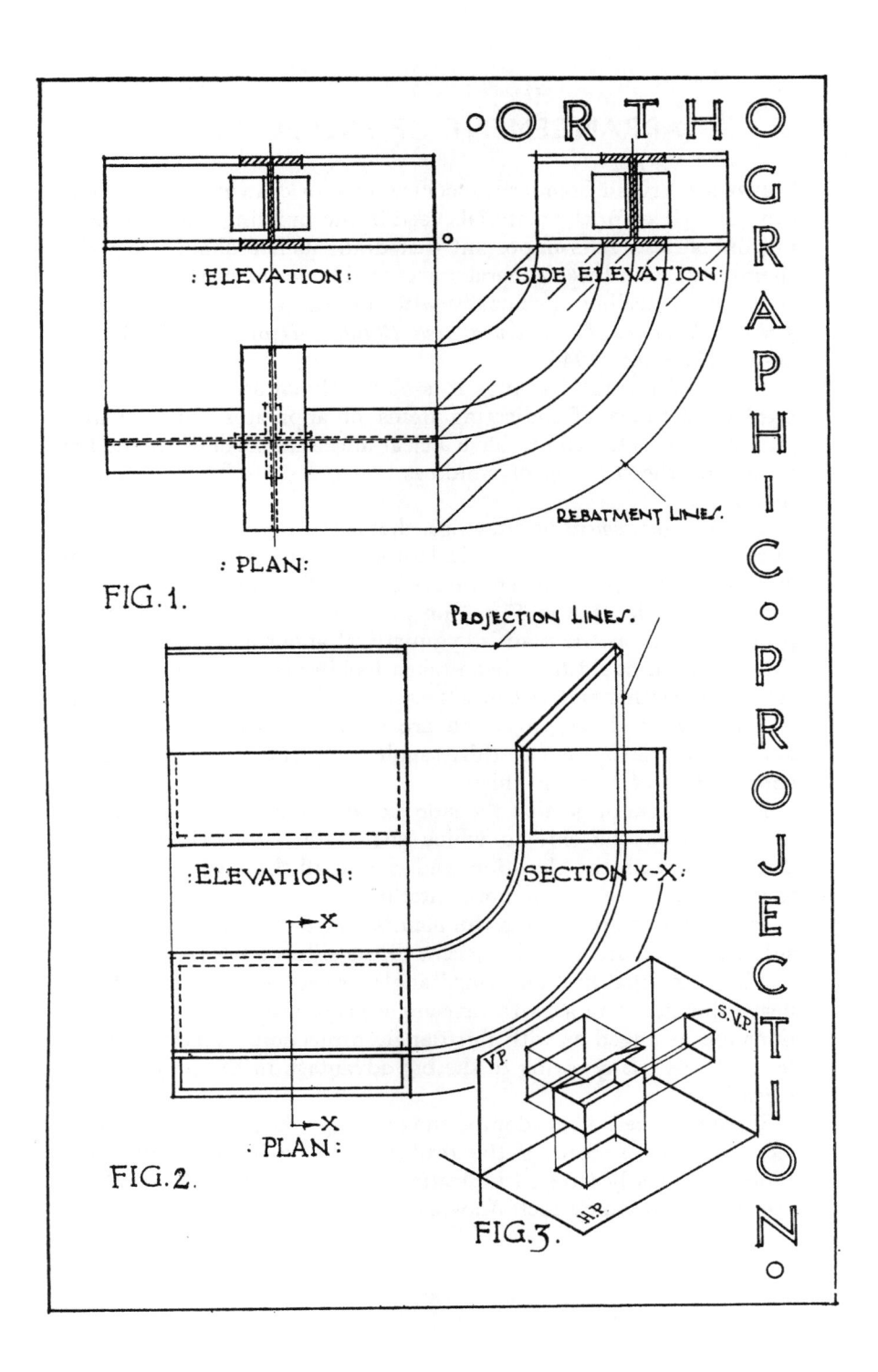

FIG. 1.

: ELEVATION :   : SIDE ELEVATION :

: PLAN :

REBATMENT LINES.

FIG. 2.

PROJECTION LINES.

: ELEVATION :   : SECTION X-X :

· PLAN ·

FIG. 3.

V.P.   S.V.P.   H.P.

ARRANGEMENT OF PROJECTIONS

THERE are British Standard Specifications for drawing-office procedure as well as for the materials used in the building and engineering industries. It is important that draughtsmen should acquaint themselves with these generally accepted codes of procedure. The publication dealing specifically with architecture and building is *British Standard Architectural and Building Drawing Office Procedure, B.S. 1192 : 1944.*

For architectural drawings it is obviously an advantage to standardize methods of indicating items of apparatus, sinks, baths, gulleys, traps, etc., and to have a clear and unmistakable method of indicating the hanging of windows and the direction of swing of doors.

In the *arrangement* of drawings, that is the position of elevations in relation to the plan, a certain latitude is usual and permitted for architects, though engineers invariably use first angle projection.

The example used in Fig. 1 on page 41 is drawn in first angle projection but as the plan is symmetrical about a vertical axis the side elevation could have been taken looking in either direction or referring to either end, as could the section X–X in Fig. 2 on page 41.

Three methods of projection are shown opposite: Fig. 1, first angle projection; Fig. 2, third angle projection, and Fig. 3, a combination of first and third.

In first angle projection the side elevations are placed at the end farthest from the side to which they refer; the plan is placed directly beneath the elevation, and as with all the other projections the elevations are on one horizontal line.

In third angle projection the plan is above the elevation and the side elevations are placed adjacent to the sides to which they refer.

The combined first and third angle projection has the plan beneath the elevation as with first angle projection, whilst the elevations are arranged as with third angle projection. This method is favoured by many owing to the big advantage in the projecting of elevations.

Whatever method is adopted the drawings should be easy to project and easy to read; if the reader could be in any doubt they should be clearly lettered indicating the direction of the view and the surface to which each drawing refers.

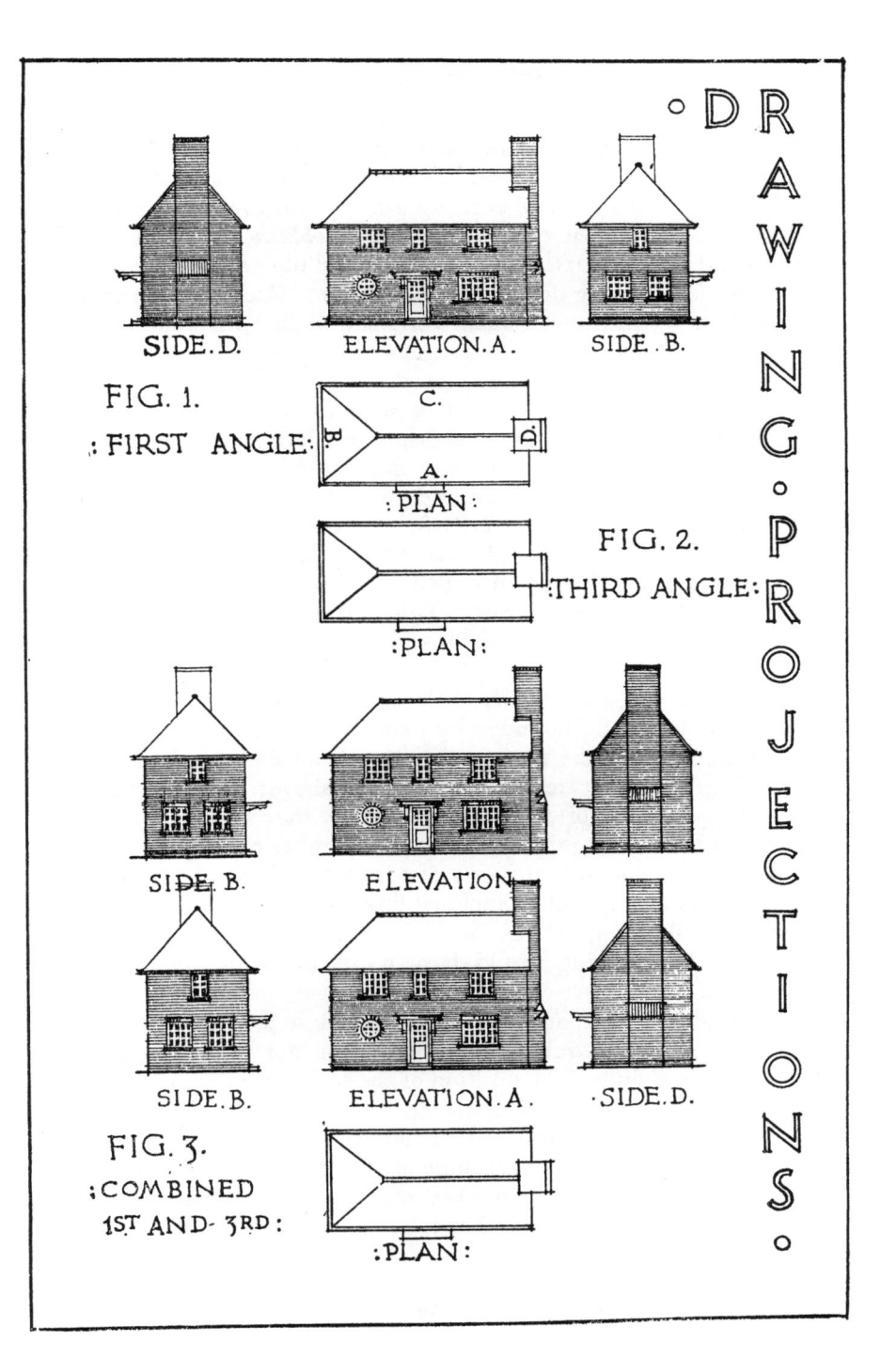

DRAWING PROJECTIONS

SIDE. D.　　ELEVATION. A.　　SIDE. B.

FIG. 1.
:FIRST ANGLE:

:PLAN:

FIG. 2.
:THIRD ANGLE:

:PLAN:

SIDE. B.　　ELEVATION

SIDE. B.　　ELEVATION. A.　　SIDE. D.

FIG. 3.
:COMBINED 1ST AND 3RD:

:PLAN:

ISOMETRIC PROJECTION

PICTORIAL PROJECTIONS are methods of presenting a solid on a plane surface as in a picture. The obvious advantage of this type of drawing is that several surfaces are exposed to view at the same time. The principal types are *isometric, oblique,* and *axonometric.*

ISOMETRIC PROJECTION is in most general use and is a convenient method for showing details of construction. The general principle is that all vertical lines remain vertical and all horizontal lines are drawn at an angle of 30°. Measurements are taken on these vertical and 30° lines but cannot be taken on an inclined line.

Conventional isometric projection such as this gives, however, a slight distortion to the dimensions, particularly on large drawings. This is not a big disadvantage as the distortion is only small and for ordinary details of construction unnoticeable.

True isometric projection calls for the use of an isometric scale which is illustrated in Fig. 4. The true or actual scale is set up on a 45° line and the divisions dropped vertically on to a 30° line which becomes the isometric scale. This "true method" is not often used and in practice the normal scale is used for both vertical and 30° lines.

It is essential, particularly with complicated details, to consider the figure as a solid bounded by plane vertical and horizontal surfaces first and to take all measurements from these surfaces.

In Fig. 1 the first treatment of Fig. 3 is shown. There is no need to make one solid of this simple example, the two girder sections conveniently lending themselves to treatment as rectangular prisms.

In Figs. 5 and 6 two views of a box are shown and the open lid gives an example of an inclined line. These are drawings of the same box shown previously in orthographic projection and reference to this drawing would have to be made in preparing the isometric.

The lid is open at an angle of 45°, which in isometric projection becomes 30°, consequently neither the top nor the under surface of the lid is visible. The position of the top edge is found by treating it as the apex of a triangle.

A triangle in isometric projection is shown in Fig. 7. It is obtained by drawing the base line at 30° and the vertical height on a vertical line. It will be noticed that neither of the sloping sides measures its real length as shown in the original figure.

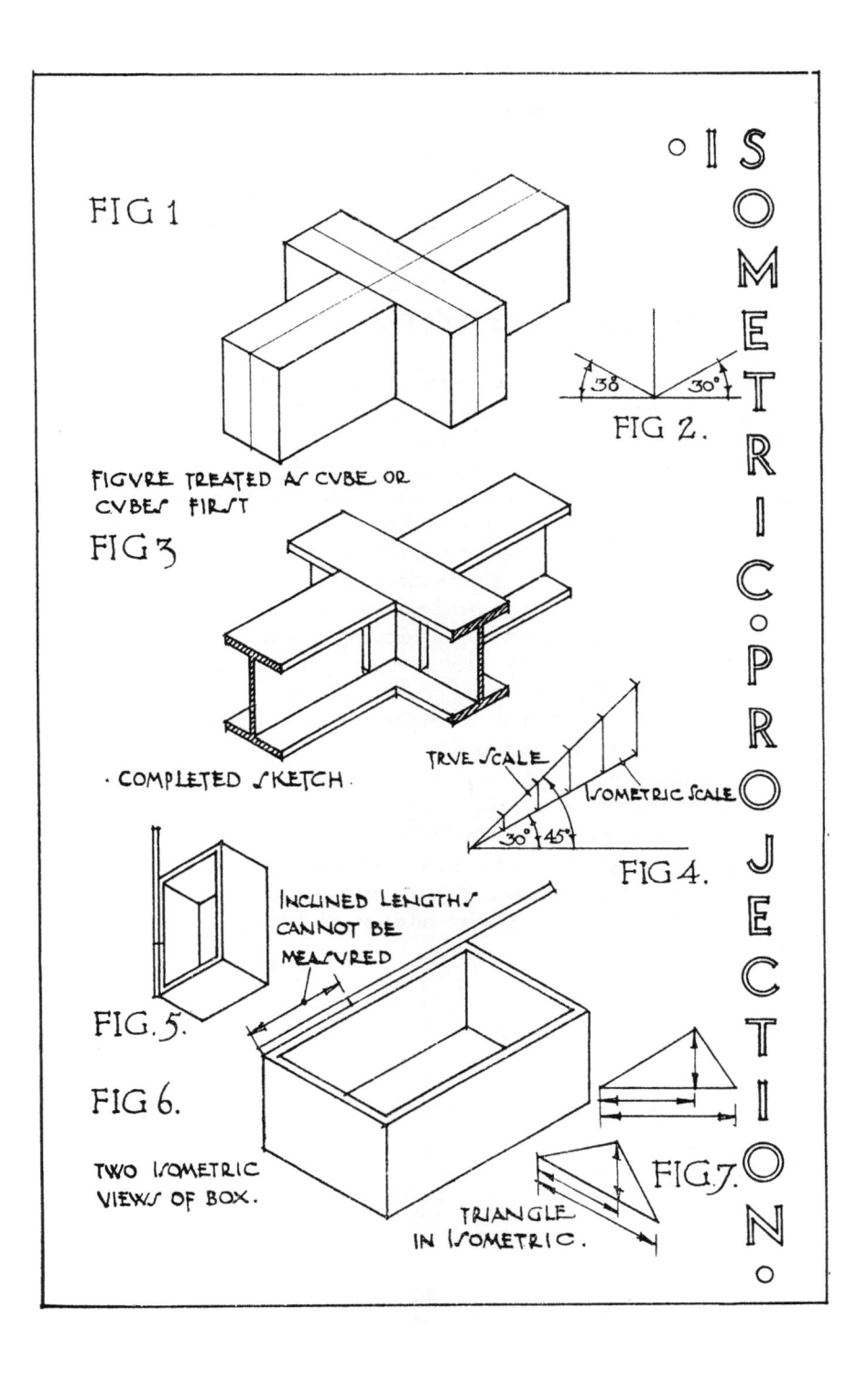

ISOMETRIC PROJECTION

FIG 1

FIG 2.

FIGURE TREATED AS CUBE OR CUBES FIRST

FIG 3

COMPLETED SKETCH

TRUE SCALE
ISOMETRIC SCALE

FIG 4.

INCLINED LENGTHS CANNOT BE MEASURED

FIG 5.

FIG 6.

TWO ISOMETRIC VIEWS OF BOX.

FIG 7.

TRIANGLE IN ISOMETRIC.

DETAILS IN ISOMETRIC PROJECTION

THREE examples are shown here of constructional details in isometric projection. Figs. 1 and 2 are of a pair of single-haunched tenons on the bottom rail of a panelled door.

The drawing should be commenced as Fig. 1. First draw the solid piece of timber, mark out the thickness and depths of the tenon and haunch, building up the finished drawing by working back from the surfaces forming the original solid.

Always use plenty of construction lines, drawn faintly, and do not rub any of them out until the sketch is complete.

Figs. 3 and 4 are stages in the working of a moulded section out of a block of stone. The processes, which consist of a series of checks and chamfers, are easily drawn and show clearly a procedure that would be quite difficult to explain or to illustrate by means of plans, elevations, and sections.

A dovetailed halving joint is shown in Figs. 5 and 6, first assembled and then with the two pieces of timber separated.

These drawings are of simple details, but illustrate how convenient this projection is for this type of drawing, either to show working processes, complete details, or, as in the case of the dovetailed halving joint, the relative position of details before and after fixing.

FIG. 1. FIG. 2.
: TENONS IN DOOR RAIL :

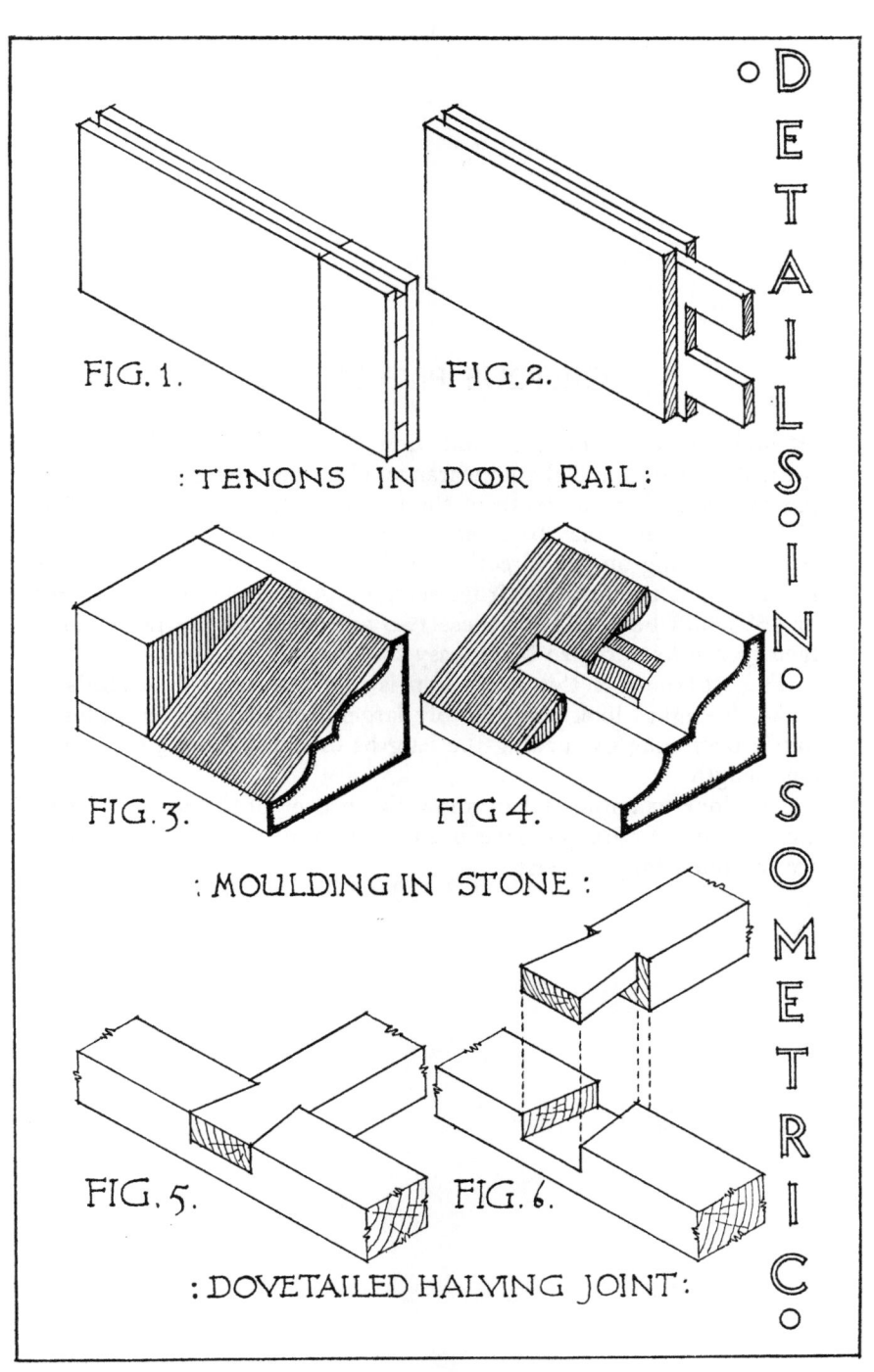

FIG. 3. FIG. 4.
: MOULDING IN STONE :

FIG. 5. FIG. 6.
: DOVETAILED HALVING JOINT :

DETAILS・IN・ISOMETRIC

D

OBLIQUE PROJECTION

IN oblique projection the front view of the object is drawn as a normal elevation and the appearance of solidity given to the drawing by projecting the depth of the figure at an angle of 30° or 45°.

The front view may be a section, a front elevation or side elevation depending on the direction of the view. In Figs. 1 and 2 the front view is the normal girder section. Any angle other than 30° or 45° could be used, but these two are usual, giving fairly satisfactory results and also being easy to construct.

The weakness of this projection is that the depth line tends to make the figure look considerably larger than it really is. This is partly overcome by making the lengths on these lines $\frac{1}{2}$ or $\frac{2}{3}$ their true length.

The four examples taken show the use of both angles. These are the same figures as have been used in all the projections and are intended for comparison.

OBLIQUE PROJECTION

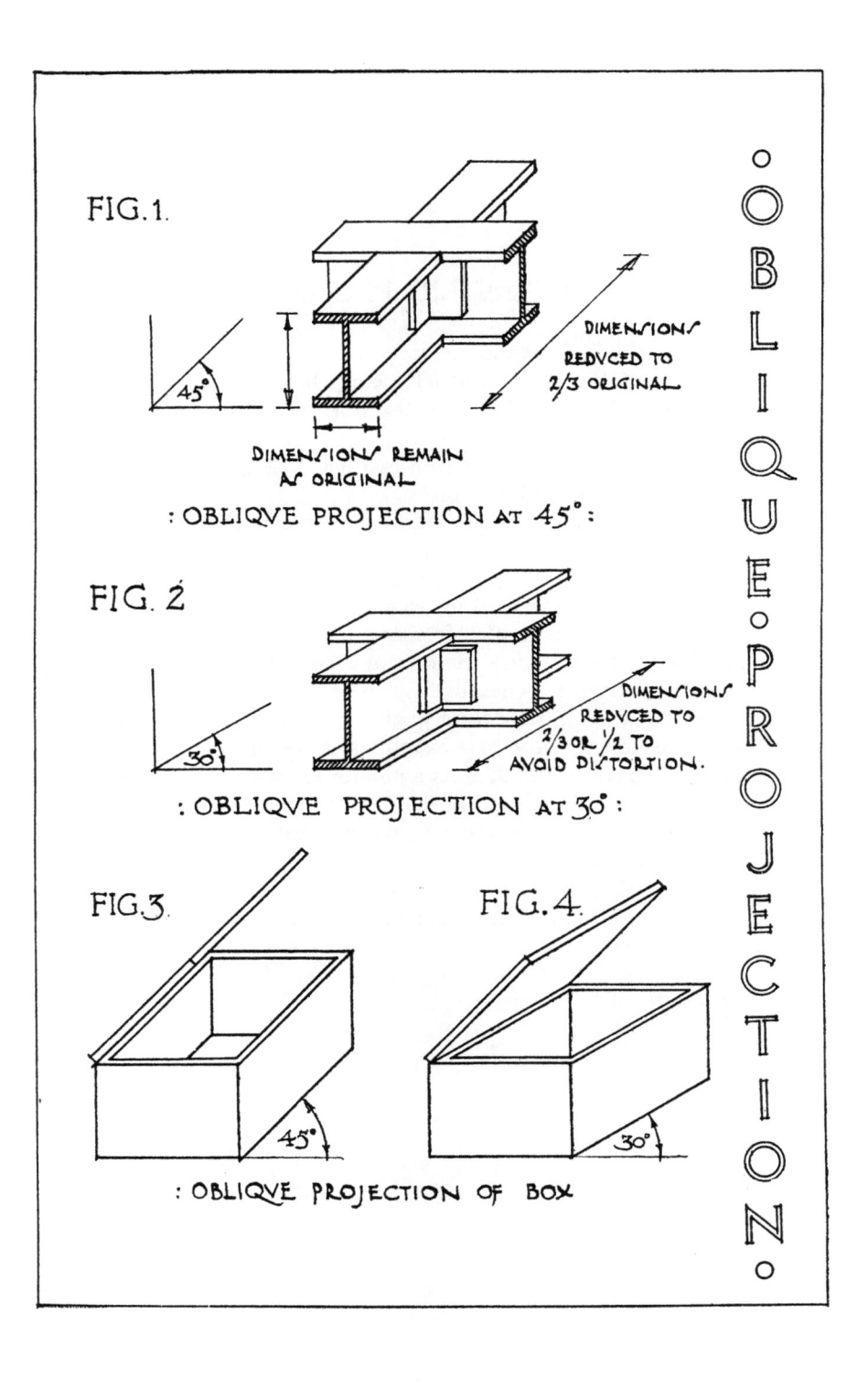

FIG. 1.

45°

DIMENSIONS REMAIN AS ORIGINAL

DIMENSIONS REDUCED TO 2/3 ORIGINAL

: OBLIQUE PROJECTION AT 45° :

FIG. 2

30°

DIMENSIONS REDUCED TO 2/3 OR 1/2 TO AVOID DISTORTION.

: OBLIQUE PROJECTION AT 30° :

FIG. 3.   **FIG. 4.**

45°   30°

: OBLIQUE PROJECTION OF BOX :

AXONOMETRIC PROJECTION

The angles used in axonometric projection are complementary angles, 45° and 45° or 30° and 60° being the usual combinations.

This leaves the angle between the lines 90° and makes the top view always the true plan.

The examples of this projection in Figs. 1, 2, 3, and 4 show the use of both 45°, 45° and 30°, 60°.

True lengths may be measured on all lines that were originally horizontal or vertical, but as with the other metric projections originally inclined lines do not appear in their true lengths and must be treated as described in isometric projection.

The fact that the true plan is shown makes this projection very suitable for pictorial views of interiors, positions of fitments and furniture, and layouts generally.

The examples of intersecting girders and the box with an open lid are not necessarily the best examples of the use of all these projections, but have been taken as a simple exercise to include horizontal, vertical, and inclined lines.

Comparison should be made between them and the merits and demerits of each noted.

The projection used when presenting a certain subject will depend on the circumstances and the particular surface view it is required to accentuate.

AXONOMETRIC·PROJECTION

TRUE PLANS

FIG.1.   **FIG.2.**

ACTUAL DIMENSIONS TAKEN IN BOTH DIRECTIONS.

INCLINED LENGTHS CANNOT BE MEASURED

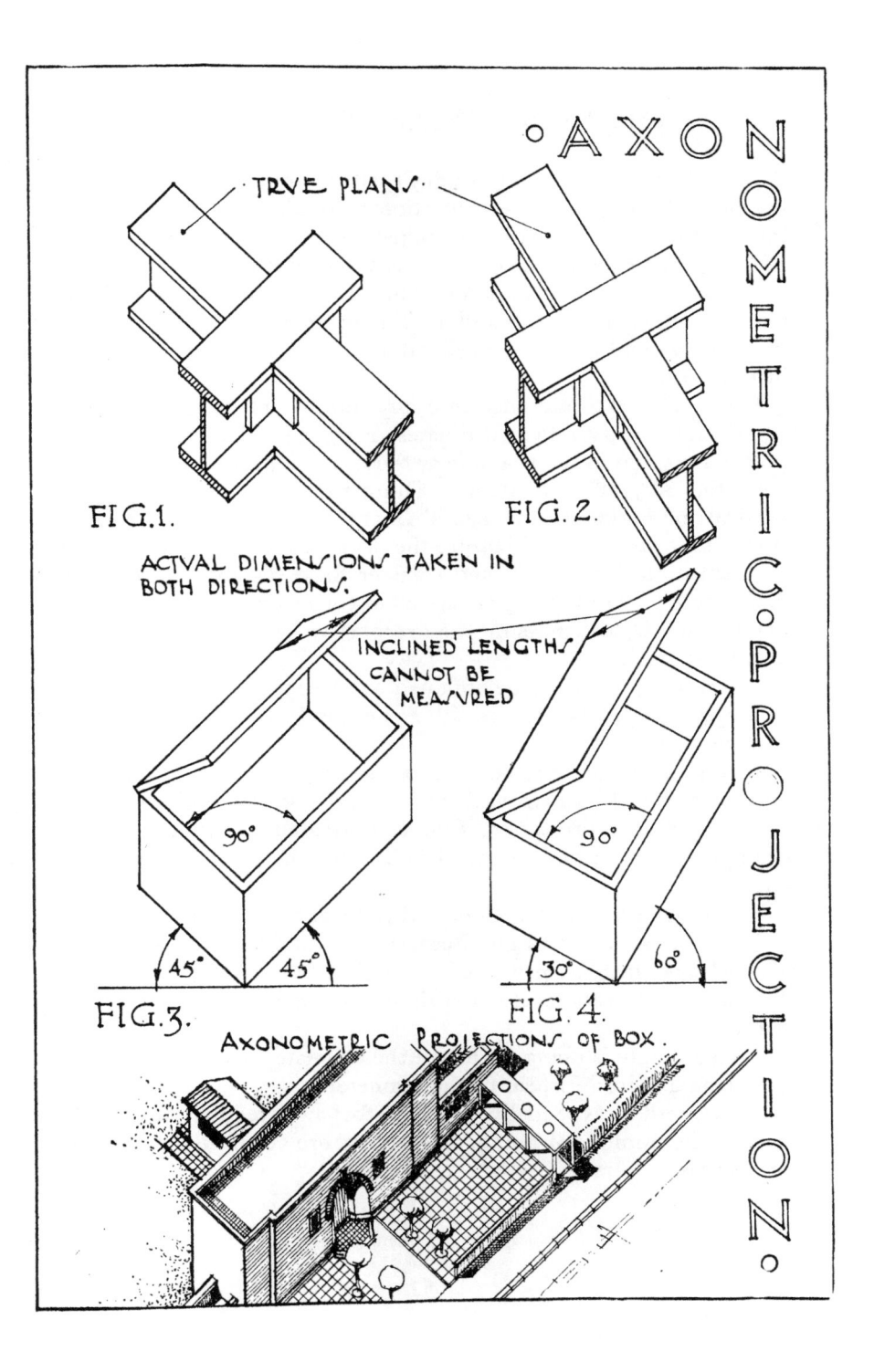

90°   90°

45° 45°   30° 60°

FIG.3.   **FIG.4.**

AXONOMETRIC PROJECTIONS OF BOX.

THE PROJECTIONS OF A CIRCLE

THE projections of a circle are dealt with more fully in that part of the book dealing with the projection of solids. On this page it is intended to deal only with the subject as far as it concerns the use of the circle or part of a circle in the usual drawing projections.

The elevation and plan of a circular disc are shown in Fig. 1 and two projections in Figs. 3 and 5. The projected curve is an ellipse in each case and is drawn freehand after locating a series of points in the curve.

Divide the circumference into any number of parts, preferably equal, and drop vertical ordinates on to the diameter. Twelve points are taken in the example as these divisions are easy to obtain with the 30°, 60° set square. Draw the lines representing the vertical and horizontal diameters AB and CD at the angles required by the new projection. Measure the distances of the ordinates from the centre on AB as at Y and transfer these distances to the new diameter AB. Draw the projections of the ordinates, making them parallel to CD, and measure their respective heights as at X and X_1 on the original.

A series of points will be obtained by this method through which a fair curve may be drawn either freehand or with the aid of a French curve.

In Fig. 2 the circle is made into a solid pierced by another and smaller circle. Elevation and section of the solid are given.

The isometric projection, Fig. 4, is treated as in the previous example, taking a new series of points for the smaller of the two circles.

The thickness may be added either by measuring on a series of 30° lines as at T, or as an alternative method the horizontal and vertical axes may be projected through to the required thickness and the back elevation set out again in the same manner as the front curve.

If an arc only is being projected the ordinates will be taken from a horizontal chord instead of the diameter.

This method is equally suitable for an irregular curve; the greater the number of points taken the more accurate the curve in projection.

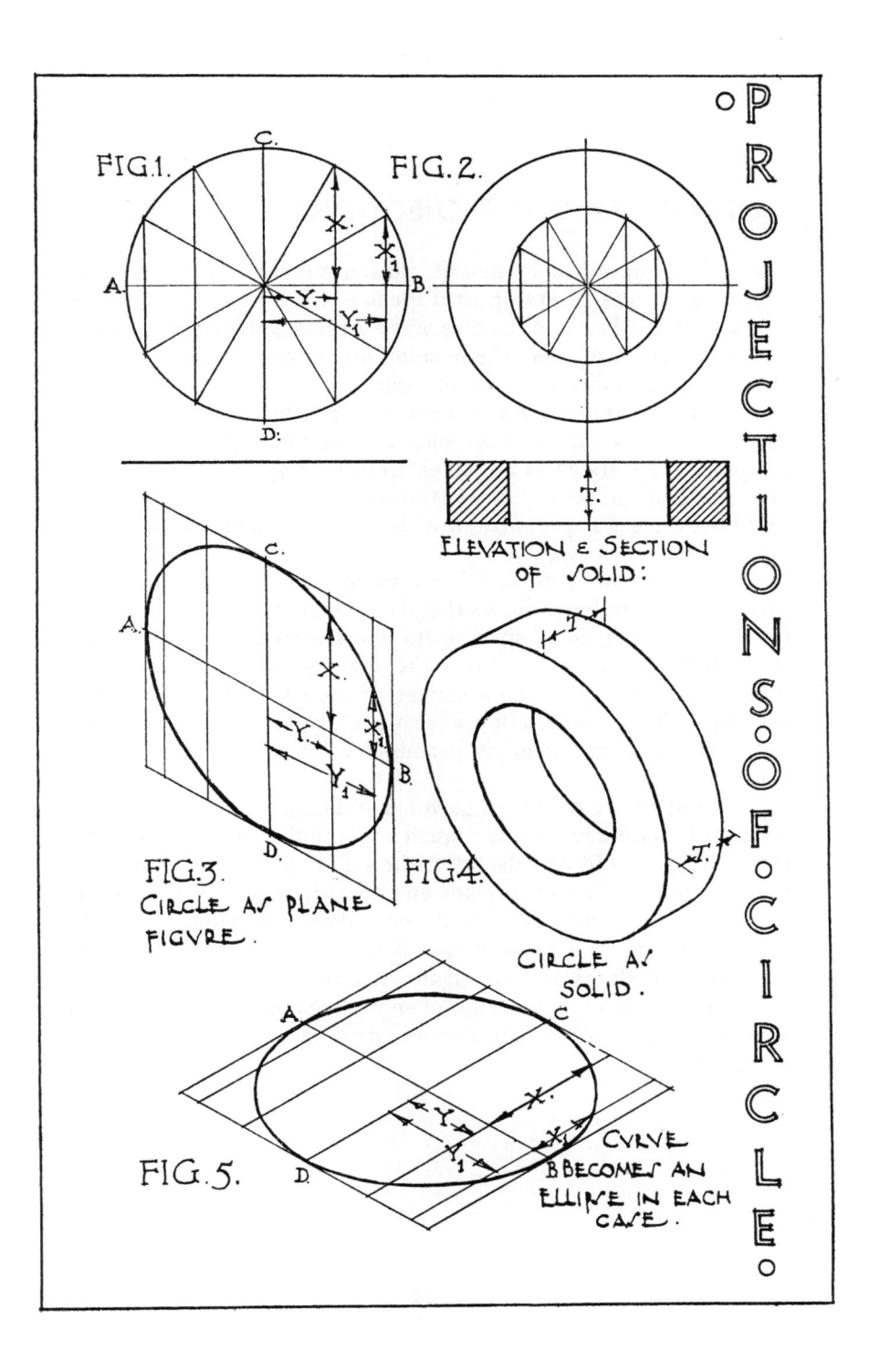

PROJECTIONS OF CIRCLE

FIG. 1.

FIG. 2.

ELEVATION & SECTION OF SOLID.

FIG. 3.
CIRCLE AS PLANE FIGURE.

FIG. 4.
CIRCLE AS SOLID.

FIG. 5.
CURVE BECOMES AN ELLIPSE IN EACH CASE.

ORTHOGRAPHIC PROJECTION: AN EXAMPLE

FOUR drawings of a memorial cloister are shown on this and succeeding pages as examples of the use of the various projections.

These examples, as in the case of the girders and box, are intended for comparison only, and the unsuitability of certain projections for this particular subject will be obvious.

It is not advisable for a student to copy these drawings at this stage, but more simple examples in construction should be attempted. The drawings are taken from a competition scheme and are intended to convey a general impression of the completed building, sizes and proportion, without giving details of construction.

The arrangement of the drawings on the opposite page is in first angle projection as far as the plan and elevation are concerned, but for convenience of spacing the cross-section has been placed beneath the plan. This makes the drawing of the section more difficult as none of the lines can be projected from either plan or elevation. The cross-section is also slightly more difficult to read but such an arrangement is permissible when the circumstances compel it.

A complete set of drawings for a building of this type would consist of: (*a*) A key plan to a small scale, indicating the position of the proposed building relative to the building line, adjoining properties, etc. (*b*) Complete plans and elevations to a scale of $\frac{1}{8}$ inch = 1 foot. (*c*) $\frac{1}{2}$ inch cale details as working drawings, consisting usually of one or two sections taken at the most advantageous positions and showing in detail the materials and methods of construction. (*d*) Full size details of mouldings and any special construction, details of which are not usual or cannot be otherwise made clear.

ORTHOGRAPHIC PROJECTION

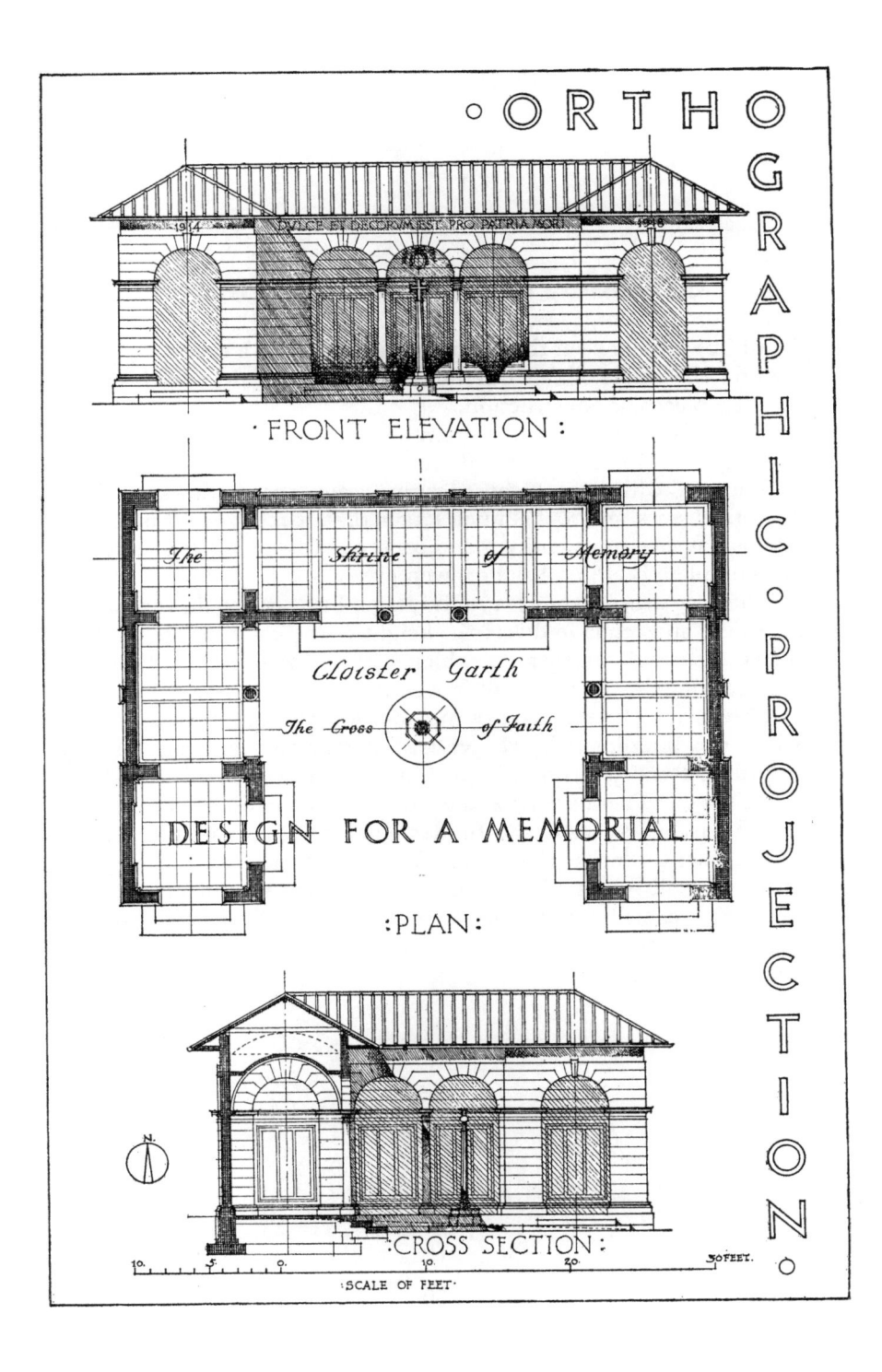

FRONT ELEVATION

The Shrine of Memory

Cloister Garth

The Cross of Faith

DESIGN FOR A MEMORIAL

PLAN

CROSS SECTION

SCALE OF FEET

ISOMETRIC PROJECTION: AN EXAMPLE

It will be readily seen that isometric is not really a suitable projection for the whole of this building although it would have been for isolated pieces of construction.

The finished work is seen from an angle from which it is most unlikely it will ever be viewed when the building is erected.

The roof with its wide overhanging eaves overshadows much of the other work and looks much larger than it really is. The elevation is not seen in its entirety and consequently a true impression of the complete building has not been given.

The east wing is shown to advantage, but here again the roof is obtrusive and tends to dwarf the height of the columns and arches. The cross is only partly seen, which is of course a disadvantage, but the arcading on the side wall is well presented.

The only really satisfactory method of pictorially presenting the whole of a building is by *perspective*. Perspective drawing is a very big subject and it is quite impossible to deal with it adequately in a few pages. There are several good books solely devoted to it, and it is advisable to study the fundamental principles of perspective rather than to devise a quick or approximate method.

As with all forms of construction, accuracy should be the student's first consideration; speed and safe approximations will follow naturally. If speed is made the first consideration, accuracy will not automatically follow.

ISOMETRIC.

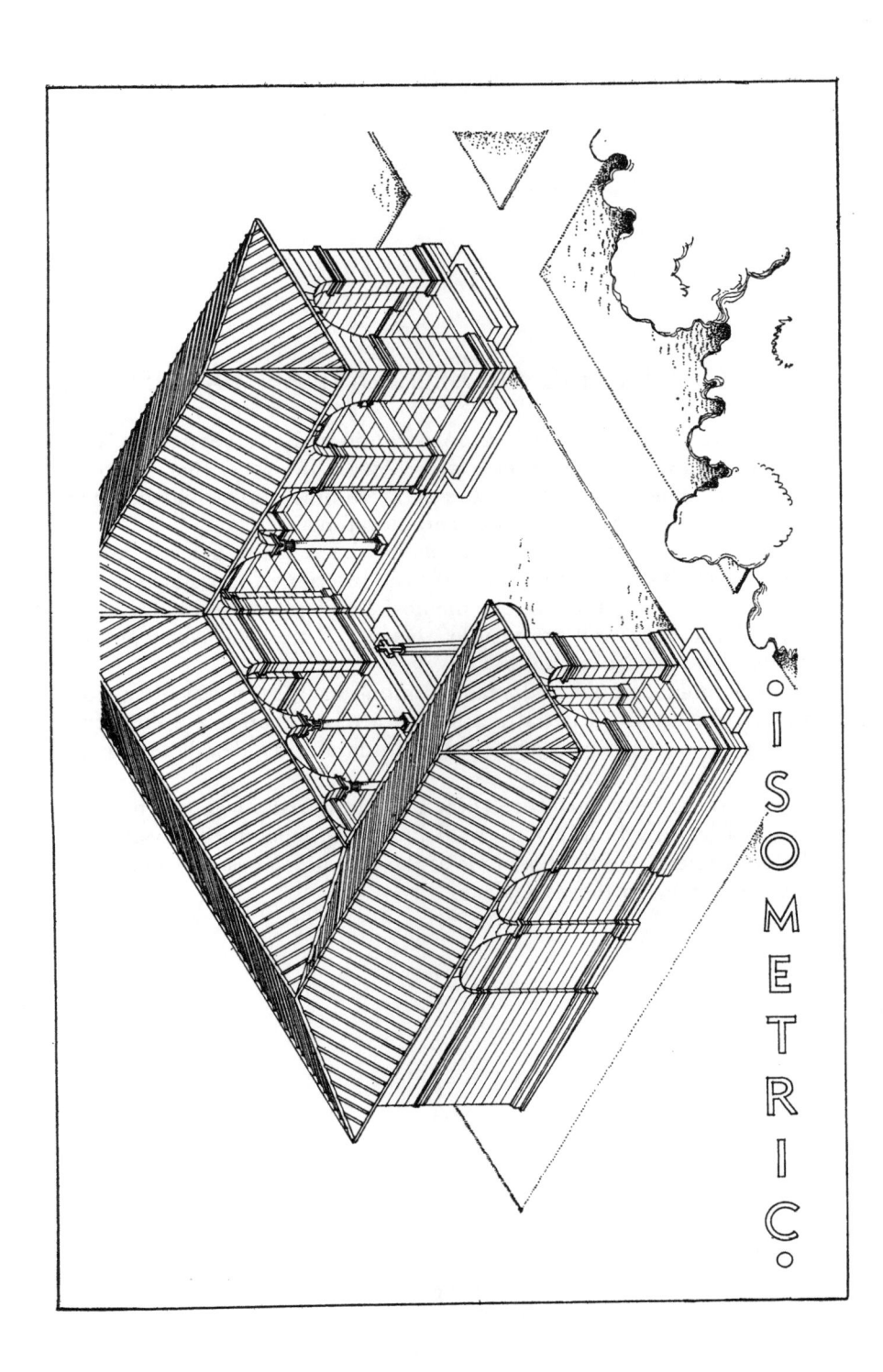

AXONOMETRIC PROJECTION: AN EXAMPLE

ALTHOUGH open to criticism on account of the steepness of the angle of view and the exaggerated view of the roof, this drawing has been made one of real value by sectioning and would be very useful to supplement the plans and elevations.

In this case a horizontal section has been taken a few courses above floor level and shows clearly the interior of the cloister.

The floor tiling and also the memorial tablets on the vertical surface of the back wall are visible. Details of the wall construction could also have been shown.

In a more complicated plan it might have been advantageous to take the sectional cut at a different level or even to step it from one level to another in order to accentuate other details either of construction or planning.

Had there been furniture (fitments, cupboards, lavatory basins, etc.) within the building it could have been shown to advantage. Before deciding on the positions of the cutting planes consideration must be given to these details.

・AXONOMETRIC・

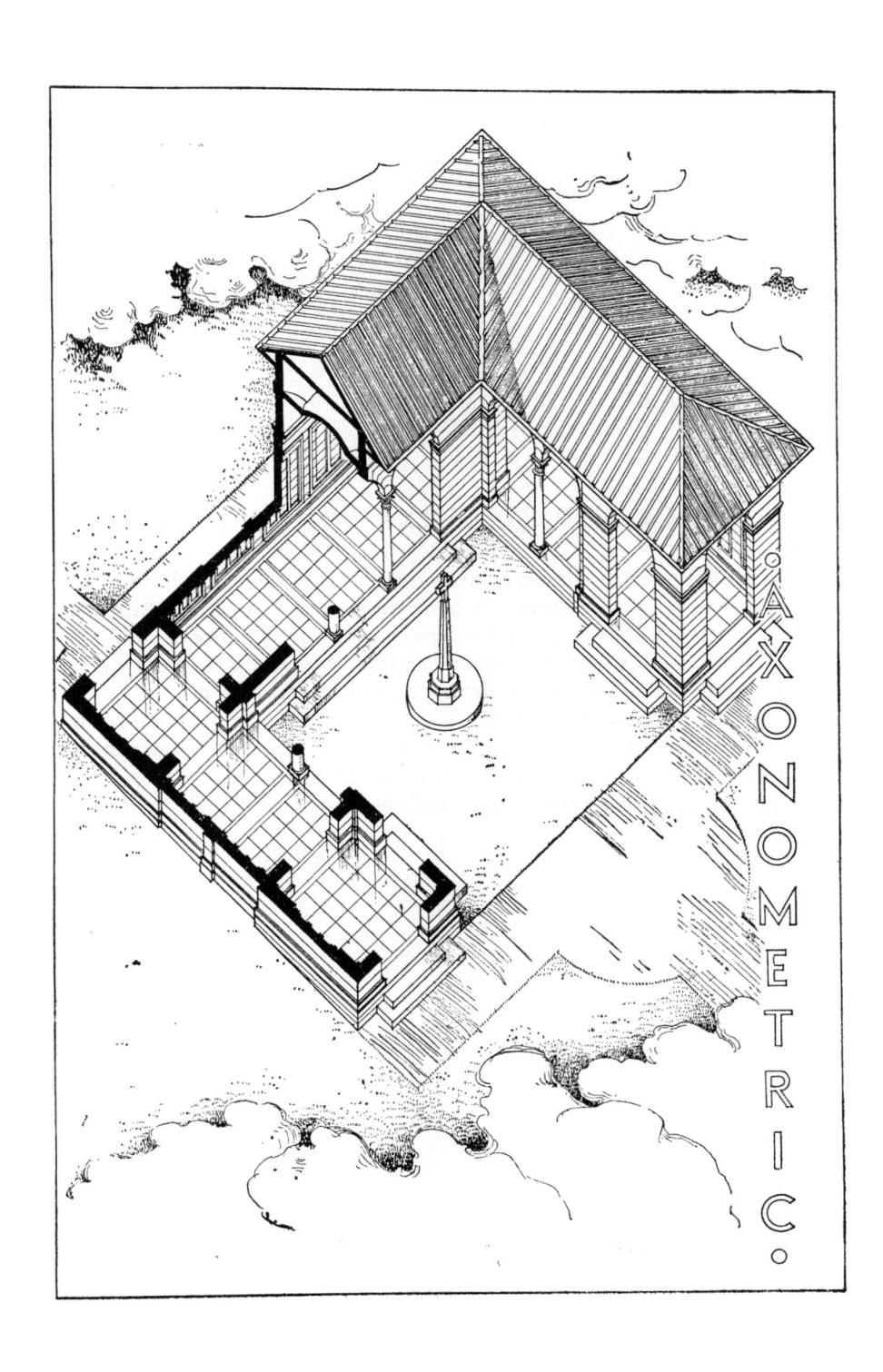

OBLIQUE PROJECTION: AN EXAMPLE

OBLIQUE projection should be confined to examples where it is desired to show the elevation or section of the object and to use the projection lines only to give depth and the appearance of solidity.

It will be seen that the complete building we are considering is not a suitable subject to be drawn in oblique projection.

The front elevations of the two wings show without distortion but the elevation as a whole is small and is made to appear even smaller by the exaggerated depth given to the wings.

Part of the elevation is hidden and the cross is also almost entirely lost to sight.

. OBLIQUE .

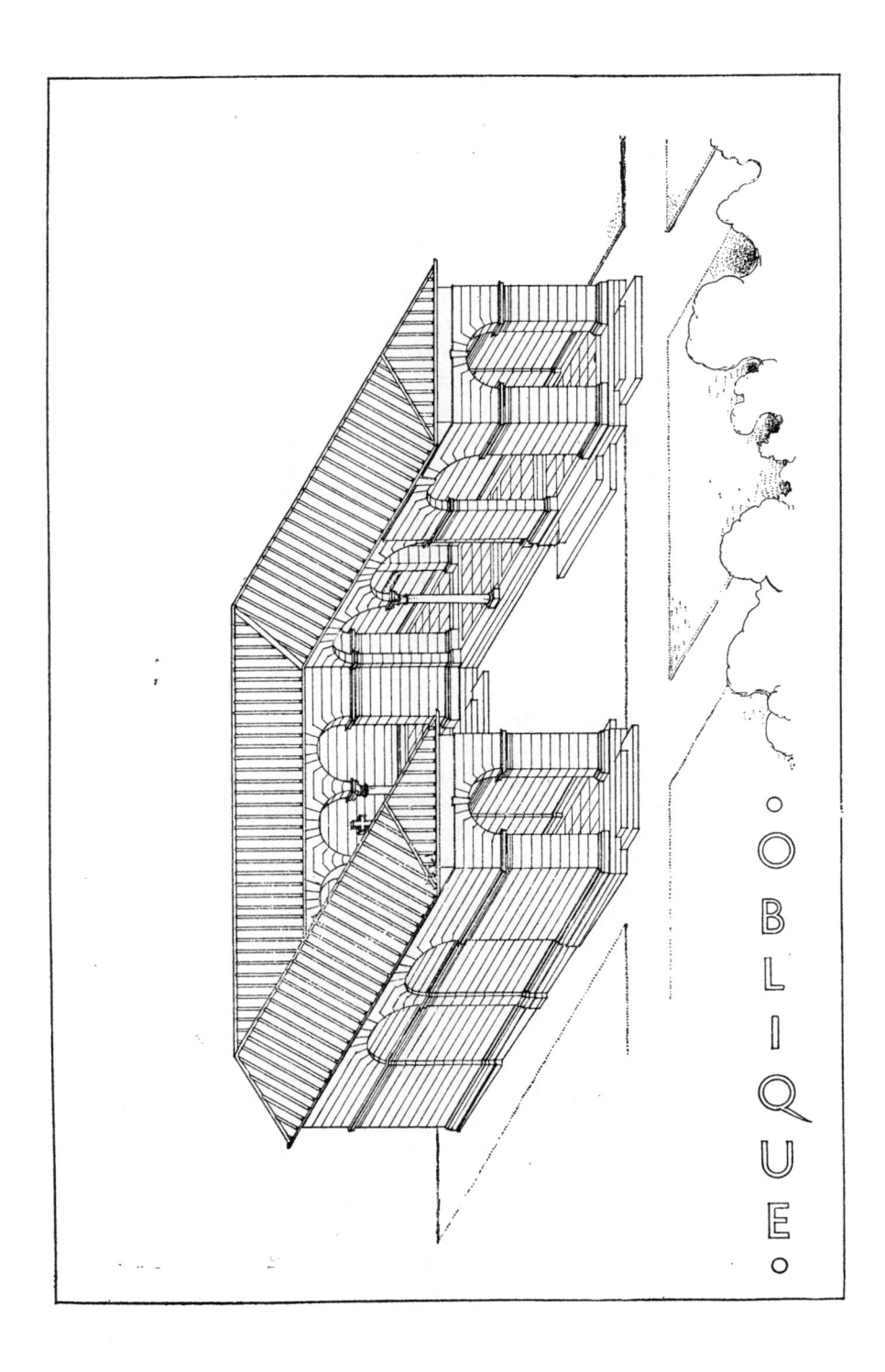

THE CIRCLE IN MOULDINGS

A MOULDED section consists of one or more members, the arrangement of which is a matter of design rather than of geometry. A section may, however, be designed and constructed entirely of geometrical members, or be drawn freehand with some supplementary geometrical construction.

Many of the members used in the classic styles of architecture, particularly the Roman, were composed of portions of circles. A selection of these constructions is given and the methods of drawing the curves need little explanation.

It is not advisable for students to use these members indiscriminately or even to design sections at all without some knowledge of the underlying principles and proportions. Mouldings should be used sparingly and reference should be made to classic examples and work of proven value.

THE OVOLO in Fig. 1 and CAVETTO in Fig. 2 are both quadrants, whilst in Fig. 3 another cavetto is shown having a similar height but less projection. The flatter curve is obtained by bisecting the chord and using as centre the intersection point between bisector and base line.

The CYMA RECTA and CYMA REVERSA in Figs. 4 and 6 have equal height and projection, the member being within a square. By dividing this square into four, by horizontal and vertical lines, centres are found for the quadrants forming the curve.

The SCOTIA in Fig. 5 is constructed on a rectangle 3 units high by 2 units wide. Divide the rectangle into six squares and the centres for drawing the quadrants are located as indicated.

Further examples of the cyma recta and cyma reversa are shown in Figs. 7, 8, and 9 and in each case the projection and the height are unequal. Draw the diagonal as AB in Fig. 7 and bisect it in point C.

With centres A, C, and B and radius AC draw arcs intersecting in D and E. These two points are the required centres. The point of separation of the curves in each case is the centre of the diagonal. A line connecting the two striking centres should pass through it and is, at that point, a normal common to both curves.

Three types of the ASTRAGAL or BEAD are shown in Fig. 10, and each is composed of a portion of one circle and therefore struck from one centre.

A similar selection of moulded members is given in that part of the book dealing with conic sections as plane figures, the curves being principally portions of the ellipse.

THE CIRCLE IN MOULDINGS

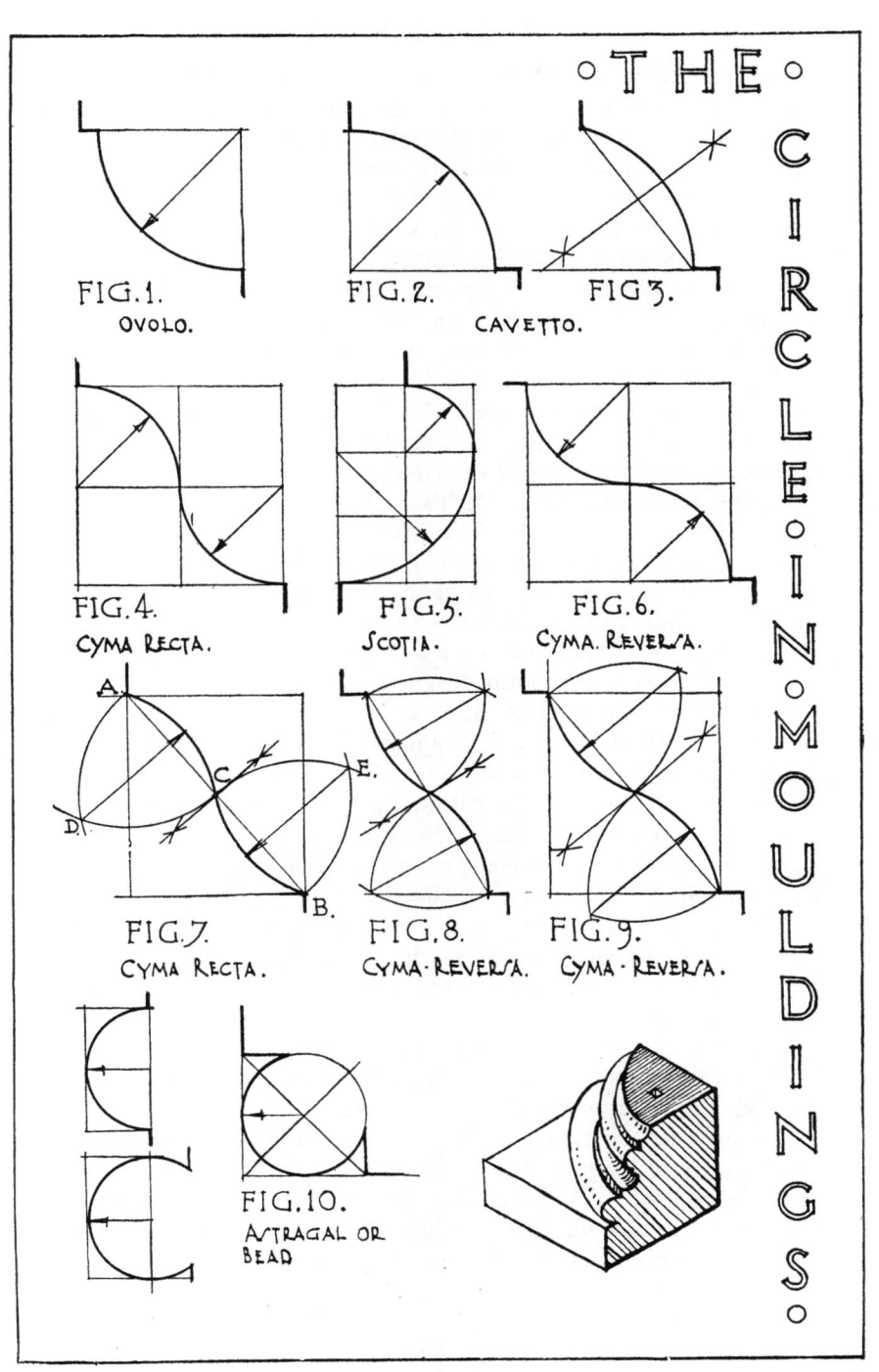

FIG. 1. OVOLO.

FIG. 2. CAVETTO.

FIG. 3.

FIG. 4. CYMA RECTA.

FIG. 5. SCOTIA.

FIG. 6. CYMA REVERSA.

FIG. 7. CYMA RECTA.

FIG. 8. CYMA REVERSA.

FIG. 9. CYMA REVERSA.

FIG. 10. ASTRAGAL OR BEAD.

ARCH CURVES

THE type of arch used depends on several factors: style of architecture, material of construction, function, and size of opening to be spanned, etc. The data supplied are usually the *span* (width of opening), *rise* (height of opening), and *type* of arch required.

The curves are set out at the soffit (underside) of the arch and are generally portions of a circle, although the ellipse is quite often used. The several examples shown opposite and on a following page are all composed of one or more portions of circles; they represent most arch curves in general use.

The OGEE ARCH in Fig. 1 is constructed on two equilateral triangles having bases equal to the span. The rise is not given but is equal to the altitude of the triangles. C_1, C_2, and C_3 are the centres, the radii are equal and the points of separation of the curves are at the intersections of the sides of the triangles.

In the second example of this arch, Fig. 2, span and rise are given. Draw the triangle DAB with base equal to the span and altitude equal to the rise, draw a semicircle on AB as diameter. The semicircle cuts the sides of the triangle in E and F; join C_1 to these points and produce the lines until they cut a horizontal line drawn from D in points C_2 and C_3, which are the centres from which the other two arcs forming the arch are struck.

The LANCET ARCH, Fig. 3, is composed of two arcs, and the centres are situated on the springing line but outside the opening.

The HORSESHOE ARCH, Fig. 4, is a little more than half a circle; the centre will therefore be situated above the springing line.

The MOORISH ARCH, Fig. 5. Here the centres will be on a line C_1C_2 at any chosen distance above the springing line; their positions are found by bisecting the chord between the rise and the span.

The STILTED ARCH, Fig. 6, is one of any shape in which the springing of the curve is above the actual springing of the arch. The vertical portion is the *stilt* and gives height without increasing the width of the opening.

The RAMPANT ARCH is one having springing points at different levels, and is the type used to support a flight of solid steps. Two methods of constructing the curve are shown in Figs. 7 and 8.

Set out the points A and B, Fig. 7, draw the horizontal AC_1, making its length equal to one quarter the span S. With centre C_1 draw the quadrant AD, join DB and bisect this chord. The bisector cuts the perpendicular, DC_1 produced, in C_2, which is the centre for the remaining portion of the arch curve.

Fig. 8 is composed of two quadrants the setting out of which is similar to the Scotia moulding.

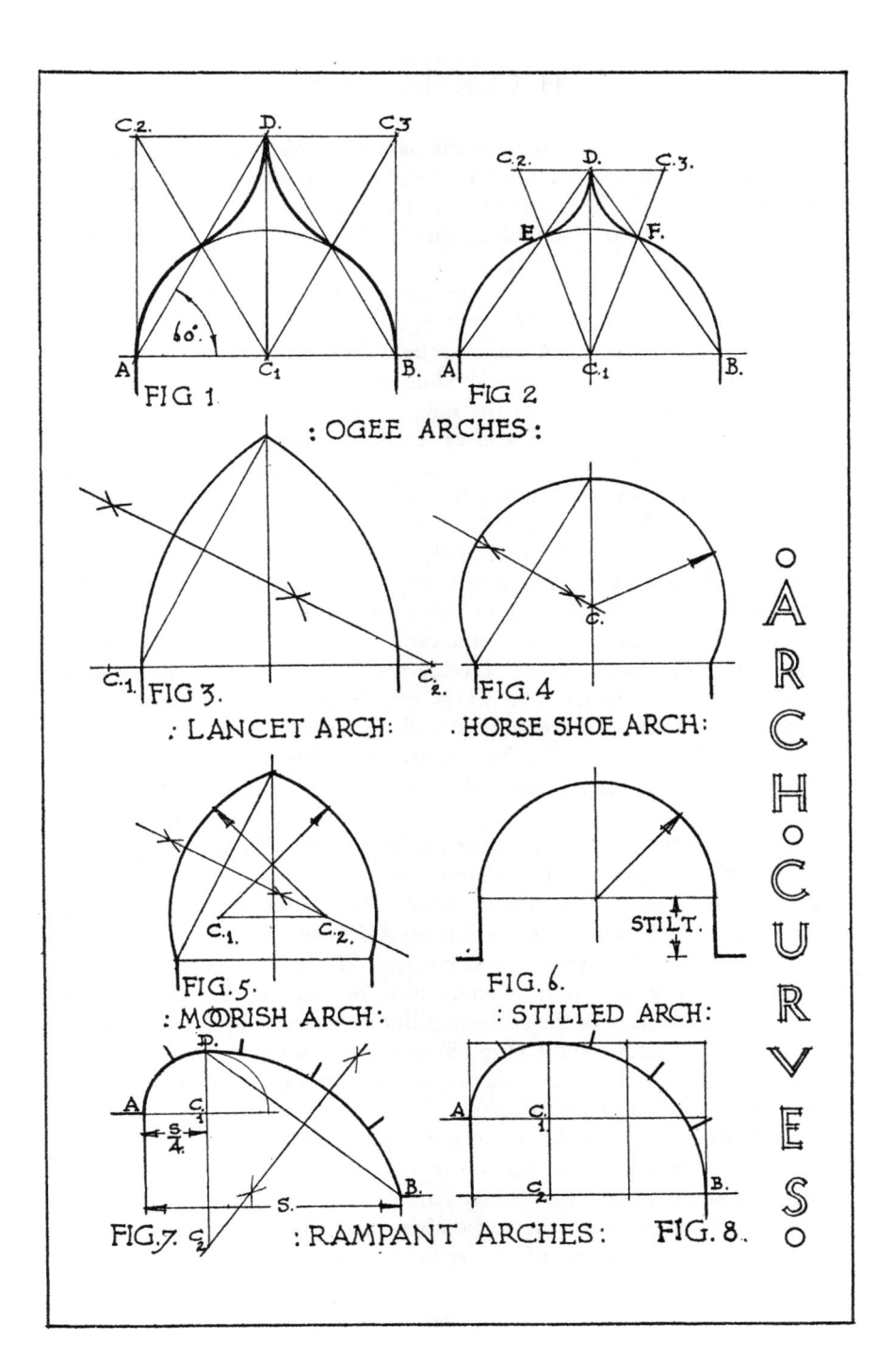

ARCH CURVES

FIG 1 / **FIG 2** : OGEE ARCHES :

FIG 3. : LANCET ARCH :

FIG 4 . HORSE SHOE ARCH :

FIG 5. : MOORISH ARCH :

FIG 6. : STILTED ARCH :

FIG 7 / **FIG 8.** : RAMPANT ARCHES :

ARCH CURVES—*continued*

The arches shown here have the parallel thickness of the arch ring added, and also the joint lines of the blocks forming the arch. It should be particularly noted that these joints are normal to the curve or curves in each case; they radiate from the centre from which the curve was struck.

The sizes of these blocks are equal and of an odd number to allow for a key block in the centre, although a central joint is permissible in the case of Gothic arches, the pointed types. It is not essential either that the blocks should be equal, it being common practice to vary the size of the key.

In the case of pointed arches if the divisions are made on the line of the soffit a key block of larger appearance results, Fig. 6. This may be partly overcome by making the divisions on a line in the centre of the face of the arch.

The SEMICIRCULAR ARCH, Fig. 1, as the name implies, is a semicircle and the SEGMENTAL ARCH, Fig. 2, any segment of a circle. An example of a brick arch of this type was shown on page 35.

A method of drawing a TUDOR ARCH is shown in Fig. 3, assuming the span only to be given. Divide the span into four equal parts and below the centre two parts construct a square. Draw the diagonals of the square which will be common normals to the curves and C, C_1, C_2, C_3 the required striking centres. Any joints below the common normal radiate to centres C and C_1, and those above to centres C_2 and C_3.

A THREE-CENTRED ARCH is shown in Fig. 4. The curve approximates to an ellipse but this is not the nearest approximation that can be obtained, as will be shown later. Having drawn the span AB and the rise OD draw a semicircle on AB. Join AD and with centre D and radius DE draw an arc cutting AD in F; bisect AF. The bisector is the common normal and the centres C_1 and C_3 are situated where it cuts the springing line and centre line respectively.

An EQUILATERAL ARCH (Fig. 5) is constructed on an equilateral triangle, the striking centres for the curve being situated at the extremities of the base or span.

A DROP ARCH, Fig. 6, is another type of pointed or Gothic arch and has less rise in proportion to the span than either the lancet or equilateral arch. The centres are on the springing line but may be anywhere between the extremities and the centre of the span. The nearer the centre of the span the nearer to a semicircle the curve becomes.

FIG 1. SEMI-CIRCULAR ARCH.

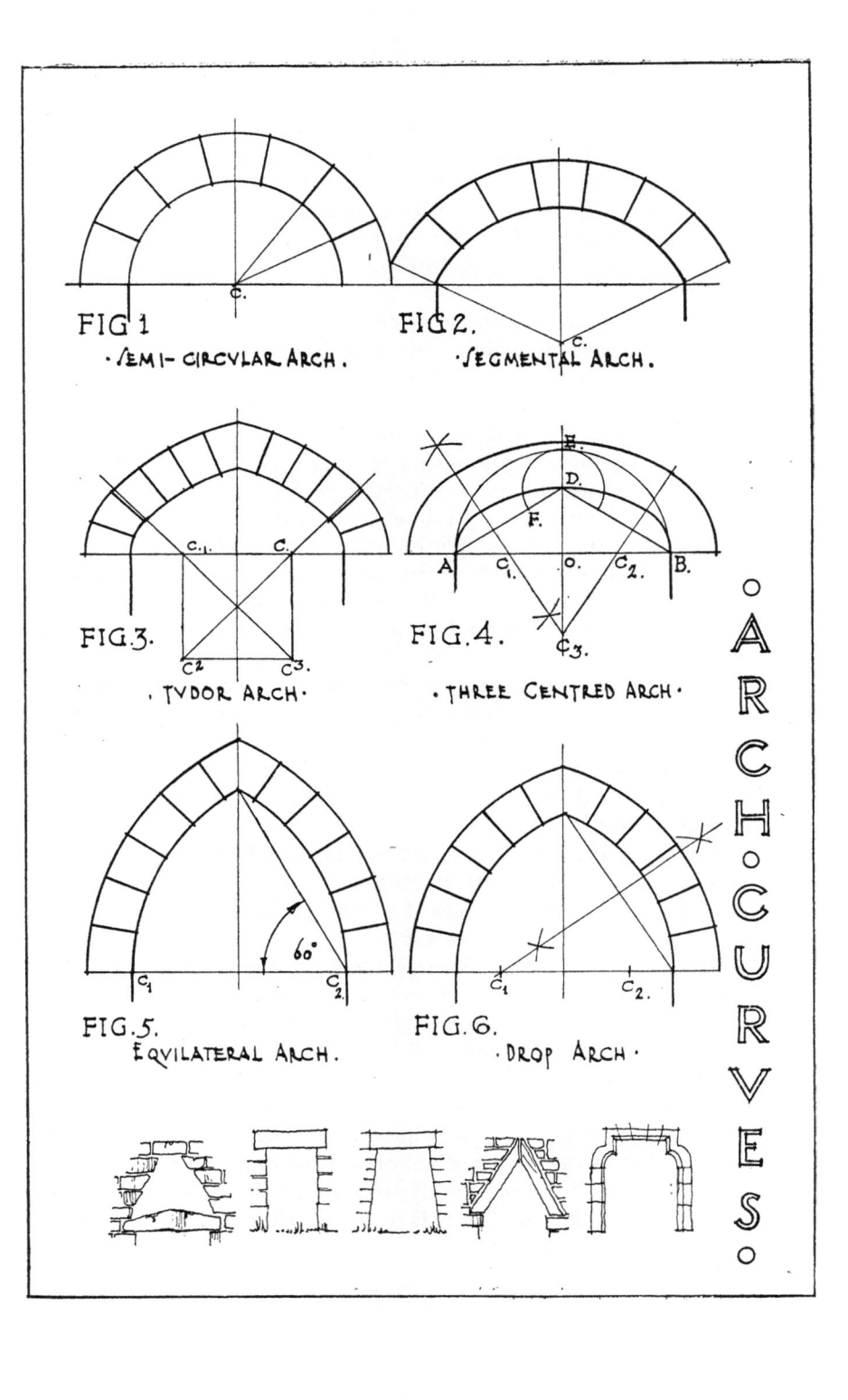

FIG 2. SEGMENTAL ARCH.

FIG 3. TUDOR ARCH.

FIG 4. THREE CENTRED ARCH.

FIG 5. EQUILATERAL ARCH.

FIG 6. DROP ARCH.

ARCH·CURVES

MOULDINGS

Mouldings can be defined as striped ribbons placed on or in a building as the design and construction demand, adjusted and proportioned to produce a harmonious and agreeable effect of light and shade.

Mouldings define and accentuate structure and form. They occur at places of change of surface, of material, and of intention of expression.

There are two distinct families of mouldings:

(a) Those derived from wooden prototypes, as in the mouldings of Egyptian and classic architecture which usually overhang or project, and

(b) Those derived from stone construction, as in medieval and Gothic architecture, which were mainly cut into the planes of stones in a series of recessed "orders".

A common fault in the design of mouldings is to make them excessive in quantity and scale. Restraint is their virtue as shown by the refinement and proportion of Greek moulded sections.

The value of mouldings with curved profiles is the production of variations of tone, creating interest and distinction. Curved sections, however, must have their boundaries controlled and separated by means of fillets and beads in order to avoid monotony of repetition.

No curved section looks well comprising more than two curved factors, and these are best when adjacently contrasted, i.e. concave and convex.

Concave mouldings reduce material and produce powerful shadows, whilst convex mouldings contain an excess of material and are robust and sturdy.

In designing mouldings repetition of identical sections is to be avoided; it is only justifiable if an even tone is desired as in the flutes of a column.

Sections of mouldings exercise a marked effect on the general character of the work upon which they are employed, and profiles must be considered with regard to the position and angle from which they will mainly be viewed.

Character in mouldings is dependent in a large measure upon the nature of the materials from which they are shaped. Mouldings in hard granite will be coarser than those in the softer limestone; similarly mouldings in deal will be larger and less refined than those in hardwood.

Shown here are a few sections of internal mouldings of the eighteenth century, in wood, which are worthy of study as being typical of one of the most refined periods of English architecture.

Mouldings are from the purely structural point of view unnecessary, and the modern tendency is to reduce their number and components or omit them entirely.

:DADO:
:CAPPING:

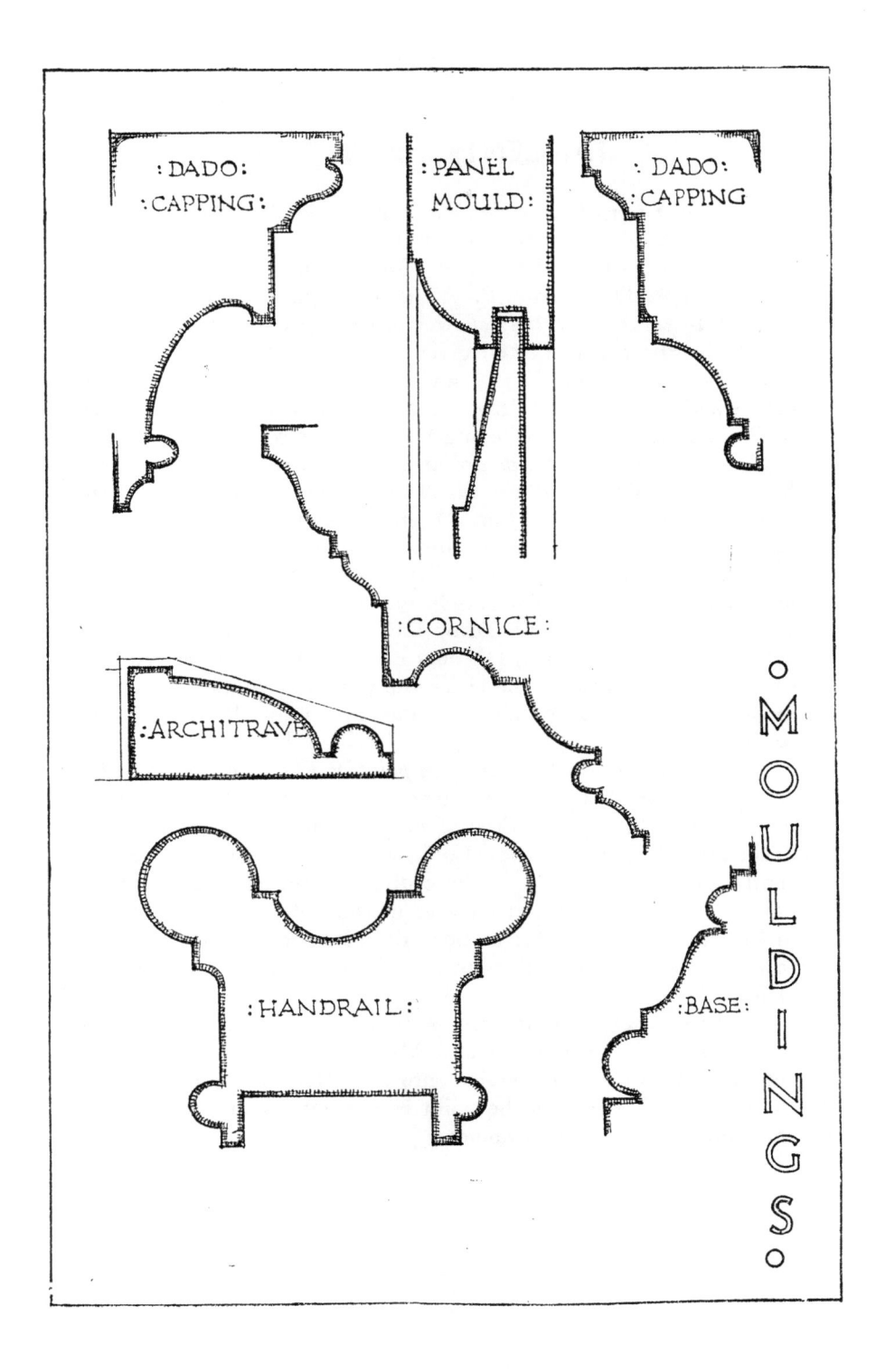

:PANEL MOULD:

:DADO:
:CAPPING:

:CORNICE:

:ARCHITRAVE:

:HANDRAIL:

:BASE:

MOULDINGS

TANGENTS TO CIRCLES

To draw a tangent to a circle at a given point of contact: Let C be the centre of the given circle and P the given point of contact, Fig. 1. Join C to P and produce. On this line, which is a normal, erect a perpendicular from P. AB is the required tangent.

To draw a tangent to a circle from a given point outside the circle: Let C be the centre of the given circle and X the given point outside it, Fig. 2. Join X to C and bisect the line in O. With O as centre and OC as radius draw a semicircle cutting the given circle in P which is the point of contact. PX is the required tangent.

To draw a tangent to an arc at a given point of contact: Let P be the given point on the arc, Fig. 3. Draw any chord PQ and bisect it, cutting the arc in R. Join RP and make the angle RPS equal to the angle RPQ. SP is the required tangent.

To draw a circle of given radius to touch two lines forming an angle: Let ABC be the given angle and AD the given radius, Fig. 4. Bisect angle ABC by BE. Draw DF parallel to AB at a distance equal to the given radius. DF cuts BE in O, which is the required centre. Before drawing the circle the points of contact P_1 and P_2 must be found by dropping perpendiculars onto AB and CB from centre O.

To draw a circle to touch a given point and be tangential to a line at a given point: Let Q be the given point and P the given point in the line AB, Fig. 5. Erect a perpendicular from P, join PQ. Make angle OQP equal angle OPQ, when O becomes the required centre and OP the radius. The problem is repeated on the same figure, using point R; and another method of solving it is shown in Fig. 6. The line QP is a chord the bisector of which will pass through the perpendicular normal from P at O, the required centre.

To draw a tangent to a circle, parallel to a given line: Let O be the centre of the given circle and AB the given line, Fig. 7. From centre O drop a perpendicular onto AB. The perpendicular cuts the circle at P, which is the point of contact. A perpendicular at this point is the required tangent.

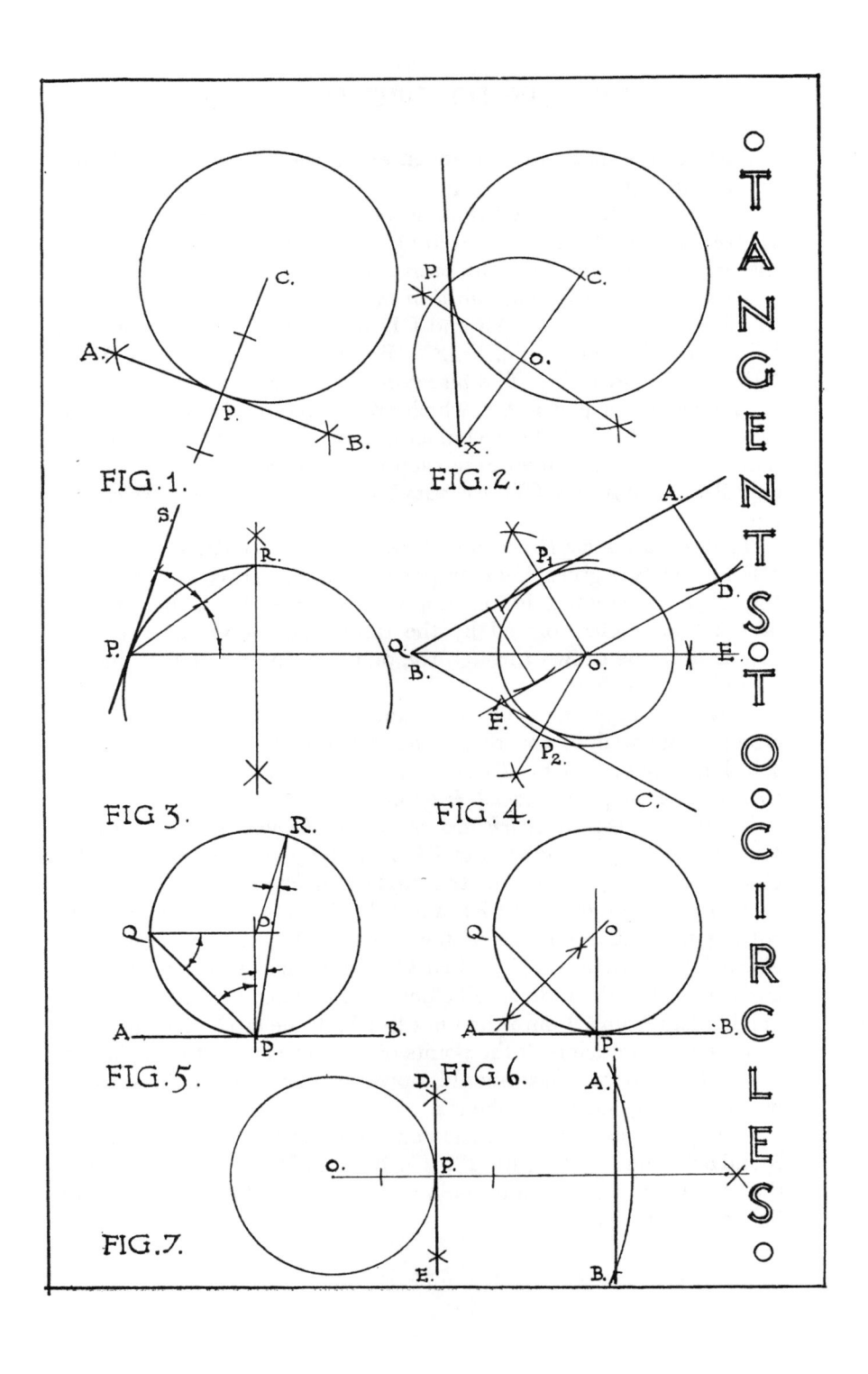

TANGENTS TO CIRCLES

FIG. 1.
FIG. 2.
FIG. 3.
FIG. 4.
FIG. 5.
FIG. 6.
FIG. 7.

TANGENTS TO CIRCLES—*continued*

These further problems on tangents give examples of circles drawn tangential to given lines.

It must be remembered that, as with examples where the circle is given and the tangent is required, the point of contact (or where there is more than one tangent, each point of contact) must be found.

To draw two circles tangential to three given straight lines, two of which are parallel: Let AB and CD be the two parallel lines and EF the third straight line, Fig. 1. Bisect each of the angles AEF, CFE, BEF, and DFE. The bisectors of one pair of angles meet in C^1 and the other pair in C^2, which are the required centres for the two circles. The points of contact are found by dropping perpendiculars from the centres onto each of the three lines as at P^1, P^2, and P^3. As AB and CD are parallel, the radii of the circles are equal.

To bisect an angle the vertex of which is inaccessible: Let AB and DE be the two given lines or portions of the arms of the angle, Fig. 2. Draw any line FG cutting the two given lines. Bisect each of the four angles formed by the three lines as in the previous problem; a line drawn through the points of intersection is the required bisector XY.

It will be seen that points C^1 and C^2 are the centres for two circles tangential to the three lines, therefore Fig. 2 is also the solution of the problem: *To draw two circles tangential to three given straight lines no pair of which is parallel.*

To draw a circle to touch two given lines forming an angle and to pass through a given point: Let EFG be the given angle and H the given point, Fig. 3. Bisect the angle and choosing any point C draw a circle tangential to EF and GF. Draw a line from H to F cutting the circumference of the circle in J, join J to centre C. Make HC^1 parallel to JC, when C^1 is the centre of the required circle and HC^1 the radius. The points of contact P and P^1 are perpendiculars from C^1 onto the lines to which the circle is tangential.

To avoid confusion all the points of contact have not been shown in the illustrations; these should, however, be found by the student when working out the problems.

In Fig. 4 is shown a practical application of tangents to a British Standard beam. The data T^1, T^2, R^1, and R^2 are obtained from the manufacturer's section book, as is the angle between web and flange, 98° in this case.

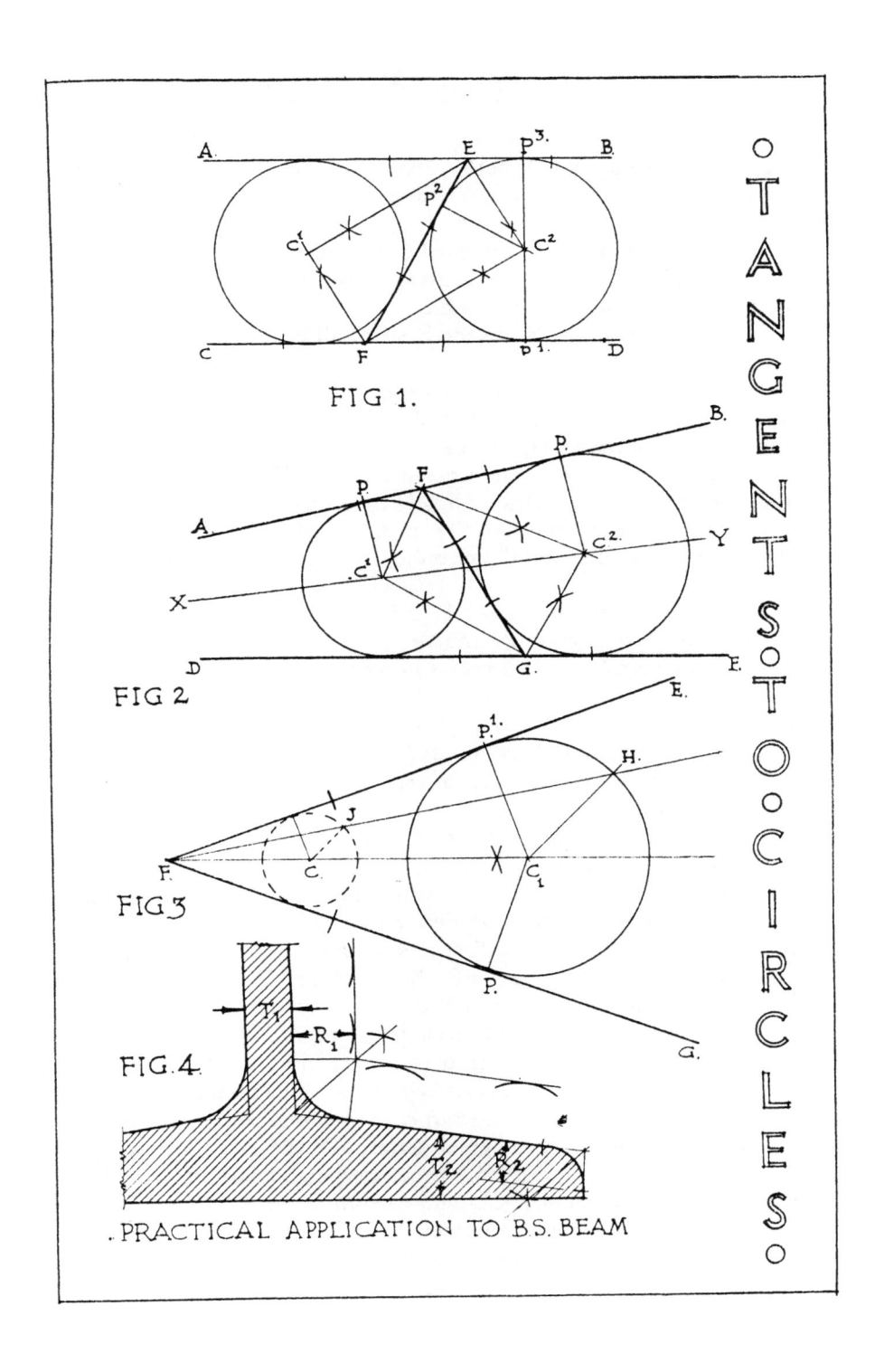

TANGENTS TO CIRCLES

FIG 1.

FIG 2.

FIG 3.

FIG. 4.

PRACTICAL APPLICATION TO B.S. BEAM

CIRCLES IN CONTACT

A TANGENT common to two or more circles is called a COMMON TANGENT and may touch both at one common point of contact or touch at separate points in each circle. Examples of each kind are shown in Fig. 1.

A COMMON NORMAL is a normal common to two or more circles. It always passes through the centres of the circles and cuts the circumference at the point of contact.

These important facts are the basis of all problems on circles in contact.

To draw two circles of given radii to touch two other given circles: Let A and B be the given circles of radii AD and BE, Fig. 2. With centre A and a radius equal to AD plus the radius of one required circle, draw an arc above the given circles. With centre B and a radius equal to BE plus the radius of the same required circle draw a second arc intersecting the first in C^1. This is the centre of the first required circle. Find the points of contact by joining C^1 to A and to B.

Repeat this operation using the other given radius to find centre C^2 and points of contact P^2 and P^3.

To draw a circle of given radius to touch two given circles, one internally and one externally: Let A and B be the centres of the two given circles and line CP the given radius, Fig. 3. With centre B and radius B plus CP draw an arc. With centre A and radius CP minus radius A draw an arc intersecting the first arc in C the required centre.

Join CB to find point of contact P and join CA and produce to find point of contact P^1.

To draw a series of circles in contact with each other and tangential to two straight lines forming an angle: Let AB and AC be the given lines and D the centre of a first given circle, Fig. 4. Drop a perpendicular from D on to AC at P. Bisect angle ADP giving point R. With R as centre and radius RP draw a semicircle which will cut the circumference of the first circle in Q and AC in S.

A perpendicular from S onto the centre line at O is the next required centre, OS is the radius and Q and S the points of contact.

To draw a circle to touch a given circle and a given straight line in a given point: Let C be the given circle, XY the given line, and A the given point in it, Fig. 5. Erect perpendiculars at C and A from XY. Perpendicular CD cuts the circle in E. Join AE and produce to B the point of contact. Join BC and produce until it cuts the perpendicular from A in C^1 the required centre.

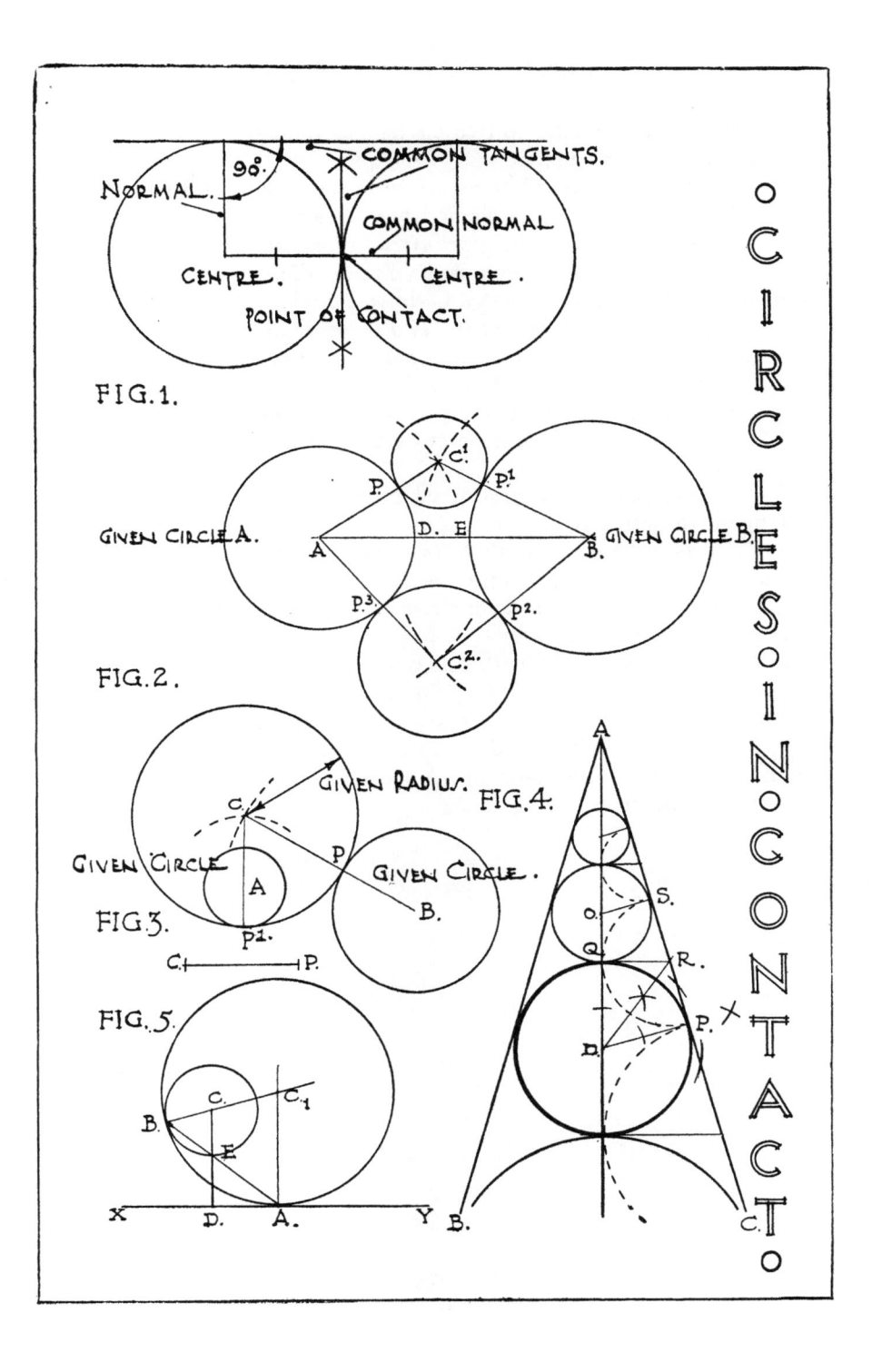

CIRCLES IN CONTACT

FIG. 1.
NORMAL. 90°. COMMON TANGENTS. COMMON NORMAL. CENTRE. CENTRE. POINT OF CONTACT.

FIG. 2.
GIVEN CIRCLE A. GIVEN CIRCLE B.
P. P¹. C¹. D. E. A. B. P³. P². C².

FIG. 3.
GIVEN CIRCLE A. GIVEN CIRCLE B. GIVEN RADIUS. C. P. P¹.

FIG. 4.
A. O. Q. S. R. D. P. B. C.

FIG. 5.
C. P. C. C₁. B. E. X. D. A. Y.

TANGENTS TO CIRCLES

AN EXTERNAL TANGENT to two circles has points of contact on the same side of both circles.

An INTERNAL TANGENT to two circles passes between them. Such internal tangents are also termed TRANSVERSE TANGENTS.

To draw tangents to two given circles of unequal radii: Let C and C_1 be the centres of the two given circles, Fig. 1. Join the centres of the circles and with centre C draw a circle the radius of which equals the *difference between* the two given radii.

Bisect CC_1 at O and with this point as centre draw a semicircle cutting the small circle in X, join XC_1. CX produced cuts the large given circle in P the first point of contact. Make C_1P_1 parallel to CP when P_1 is the second point of contact and PP_1 one required tangent.

A second external tangent may be drawn which is also a common tangent to both circles.

To draw internal tangents to two given circles: Let C and C_1 be the centres of the two given circles, Fig. 2. Join the centres C and C_1 and with centre C draw a circle the radius of which equals the *sum* of the two given radii.

Bisect CC_1 at O and with this point as centre draw a semicircle cutting the large circle in X. Join XC_1 and XC.

CX cuts the large given circle in P, the first point of contact.

Make C_1P_1 parallel to CP when P_1 is the second point of contact and PP_1 one required tangent. A second tangent may be drawn in a similar manner.

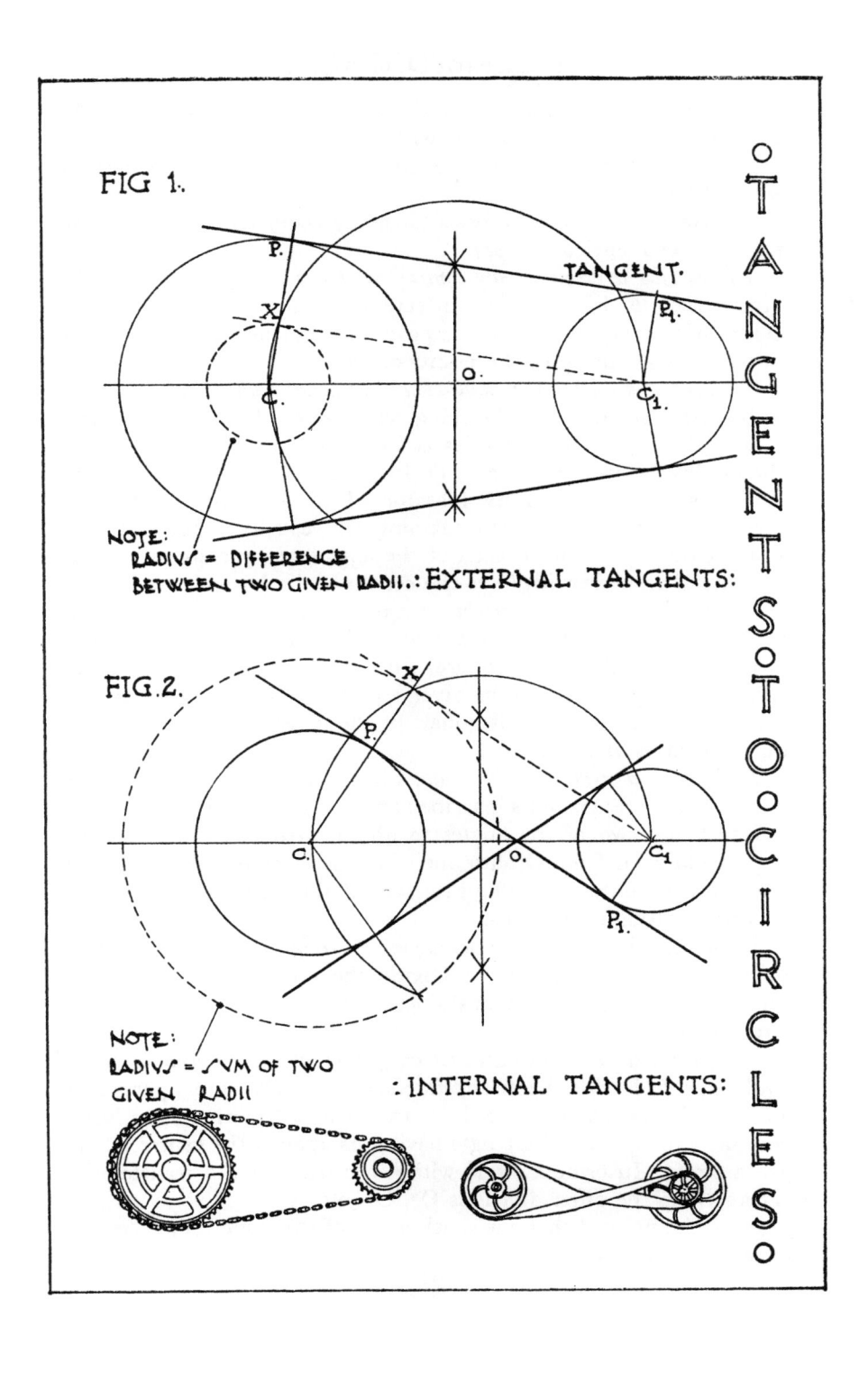

TANGENTS TO CIRCLES

FIG 1.

TANGENT.

NOTE:
RADIUS = DIFFERENCE BETWEEN TWO GIVEN RADII.

: EXTERNAL TANGENTS :

FIG. 2.

NOTE:
RADIUS = SUM OF TWO GIVEN RADII

: INTERNAL TANGENTS :

INSCRIBED CIRCLES

A FIGURE drawn within another is said to be *inscribed*. The examples opposite are circles within other regular figures; they are in contact and tangential to the sides of the given (or *circumscribed*) figures.

Inscribed circles form a basis for many geometrical patterns and for some geometrical tracery.

To inscribe a circle in an equilateral triangle: Let ABC be the given triangle, Fig. 1. The bisectors of the angles meet at the centre of the required circle. If produced they also bisect the sides of the triangle and give the points of contact.

To inscribe six circles in contact and of equal radii within an equilateral triangle: Let ABC be the given triangle, Fig. 2. The bisectors of the angles cut the sides at D, E, and F; the centres of all the circles lie on these lines. To locate them, rebisect one of the angles as at BG, cutting the bisector AF in O the first centre and giving OF the radius. The remaining centres may be found by repeating this operation in each of the angles or by constructing an equilateral triangle with O as the centre of the base.

To inscribe four circles within a square, the circles to be of equal radii, in contact with each other and with the sides of the square: Let ABCD be the given square, Fig. 3. Bisect each of the sides and draw a square within the given square. Where the sides of the inscribed square cut the diagonals of the given square are the required centres as at O.

To inscribe a circle in a hexagon, Fig. 4: The centre of a regular polygon is the centre of its inscribed and of its circumscribed circles.

Bisect any two of the angles to obtain O the required centre. A perpendicular from O onto one of the sides gives the radius and the point of contact P. The bisectors of the sides of the polygon also pass through the centre.

To inscribe five circles in a pentagon, or circles, to the number of its sides, in any polygon, Fig. 5: Divide the polygon into the same number of triangles as it has sides. In each triangle inscribe a circle as in Fig. 1.

To inscribe six circles in contact and of equal radii within a given circle, Fig. 6: Divide the circumference into twelve equal parts and at one point of division, e.g. P, draw a tangent; with this as base, complete an equilateral triangle having its apex at the centre of the given circle. Inscribe a circle within the triangle and transfer the centre O^1 to the other divisions O^2, O^3, etc.

It will be noticed that a seventh circle of the same radius can be drawn between the six.

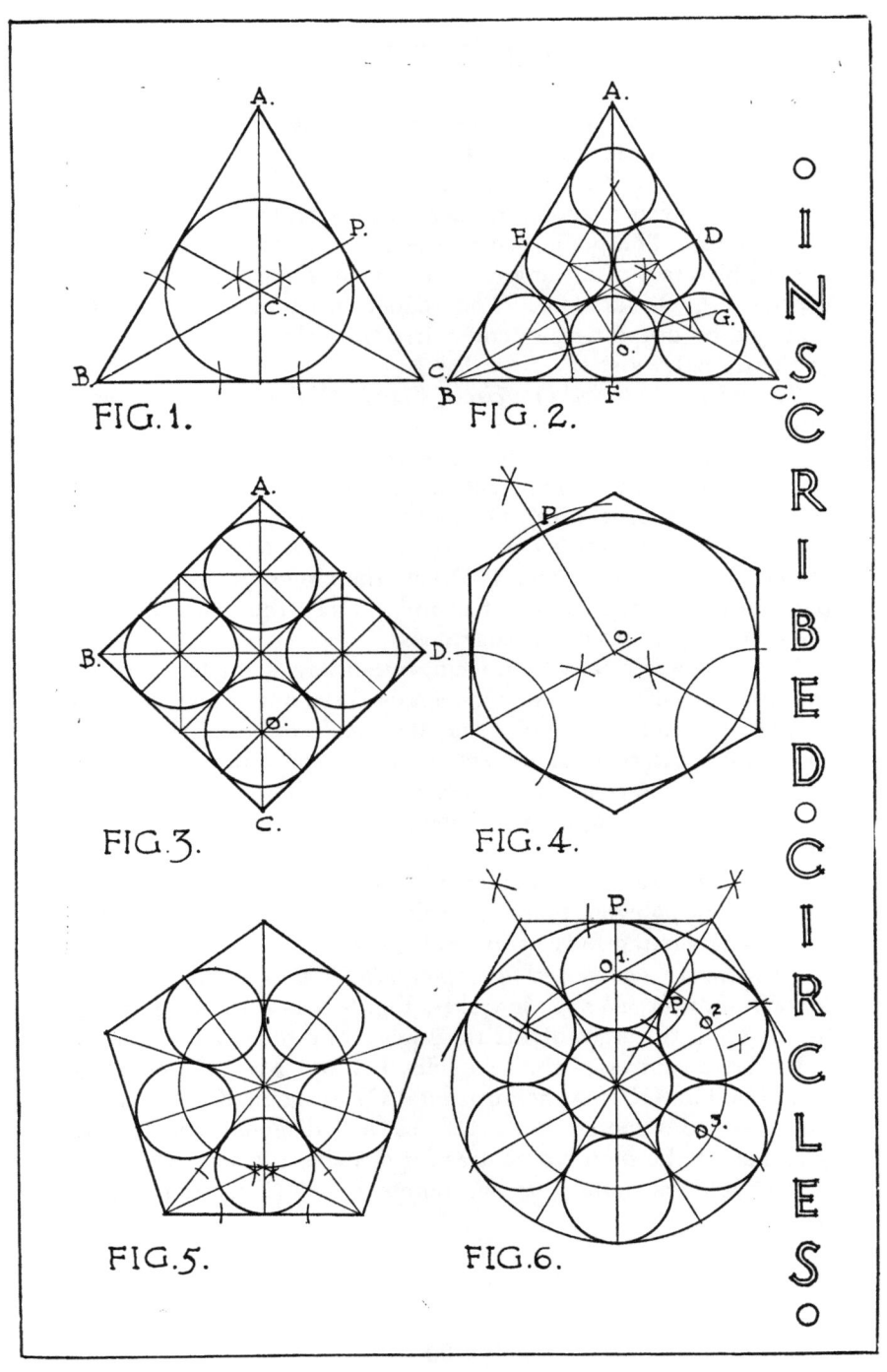

INSCRIBED CIRCLES

FIG. 1.

FIG. 2.

FIG. 3.

FIG. 4.

FIG. 5.

FIG. 6.

FOILED FIGURES

A FOIL is the circular portion between the cusps in tracery. It may be half or any other portion of a circle.

To construct a trefoil in an equilateral triangle, Fig. 1: Draw within the given triangle another triangle from the centre points of the sides. The points where the sides of the inscribed triangle cut the bisectors of the angles of the given triangle are the required centres O^1, O^2, and O^3. The radius and points of contact are obtained by dropping perpendiculars from the centres to the sides of the triangle as at P.

To construct a trefoil in a circle, Fig. 2: Divide the circumference into six equal parts. Join these points by diameters and chords forming two inscribed triangles with their medians. Each point where a side of one triangle cuts a median of the other is one of the required centres, as at O^1, O^2, and O^3.

To construct a hexafoil in a circle, each foil to be a semicircle in contact with the circle, Fig. 3: Divide the circle into twice as many parts as the required number of foils. Draw the diameters and at the extremity P of one, a tangent AB.

Bisect the angle between the tangent and the diameter, and where the bisector cuts the next diameter is the point where two of the semicircles meet. This point may be transferred from the centre of the circle to alternate diameters and the actual centres of the semicircles are at the centre of lines connecting the points.

Two similar examples are shown, a trefoil, Fig. 4, and a cinquefoil, Fig. 6.

It will be noticed that tangents to alternate diameters form a circumscribing polygon, therefore a similar procedure can be adopted when constructing foils within polygons.

To construct a quatrefoil in a square, the foils to be semicircular and tangential to the sides of the square, Fig. 5: Draw the diagonals and by bisecting the sides divide the square into four equal squares.

Find the centres of the two sides FC and BE in J and H, join JH. The line JH cuts the horizontal GE in point K. The sides of a square constructed with OK as half-diagonal cut the main diagonals at the required centres for drawing the semicircles.

It should be noted that the points J and H are not points of contact.

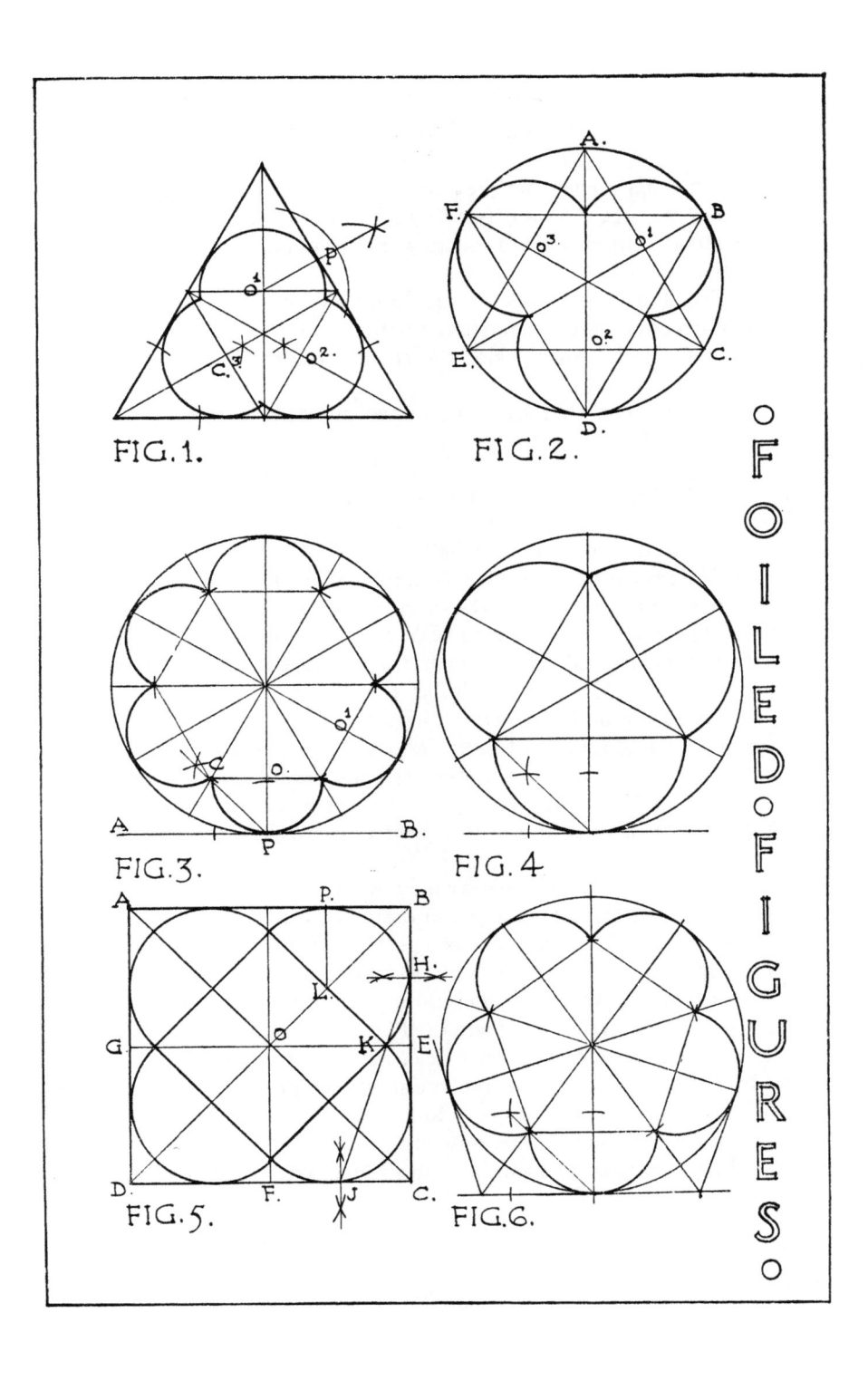

FIG. 1.

FIG. 2.

FIG. 3.

FIG. 4.

FIG. 5.

FIG. 6.

FOILED OF FIGURES

CONTINUOUS CURVES

A CONTINUOUS curve may be composed of portions of circles, or other curves, not necessarily of equal radii. The point of contact of two consecutive curves is on a common normal joining their respective centres.

In Fig. 1 a continuous curved line is shown but is of no particular practical use other than to illustrate clearly the points of contact, C, E, F, and G, also the common normals 1–2, 2–3, 3–4, and 4–5.

It will be seen that by joining two adjacent points a chord is formed, on the bisector of which a centre may be found.

The egg-shaped sewer sections in Figs. 2 and 3 are continuous curves. In Fig. 2 the diameter of the sewer is given as D. Draw two circles in contact, C_1 and C_2, making the diameter of C_1 equal to D and that of C_2 equal to half D.

Draw lines connecting the extremities of the horizontal diameters and produce until they cut the circumference of the small circle in P and Q. Bisect these chords RP and SQ. Where the bisectors meet the horizontal diameter of C_1 produced are the required centres, C_3 and C_4, for completing the curve.

A second method is shown in Fig. 3.

Draw a semicircle C_1 of radius equal to the given diameter. Find centres C_3 and C_4 by producing the horizontal diameter of C_1 a distance equal to $D/4$. With these two centres and radius equal to D draw arcs intersecting on the centre line which gives centre C_2, also the common normals and points of separation of the curves.

The arcs EG and FH are drawn from centres C_4 and C_3 and determine points G and H, giving the radius of C_2.

The example in Fig. 4 shows a series of continuous curves composed of semicircles within a circle which is itself part of the continuous curve. There is one common normal only on which, of course, the centres are situated. They are found by dividing the diameter of the circle, in this case, into eight equal parts.

In Figs. 5, 6, and 7 three moulding members are shown, the construction of which is easily followed.

Many of the mouldings and arch curves previously dealt with should be referred to again and examined as continuous curves.

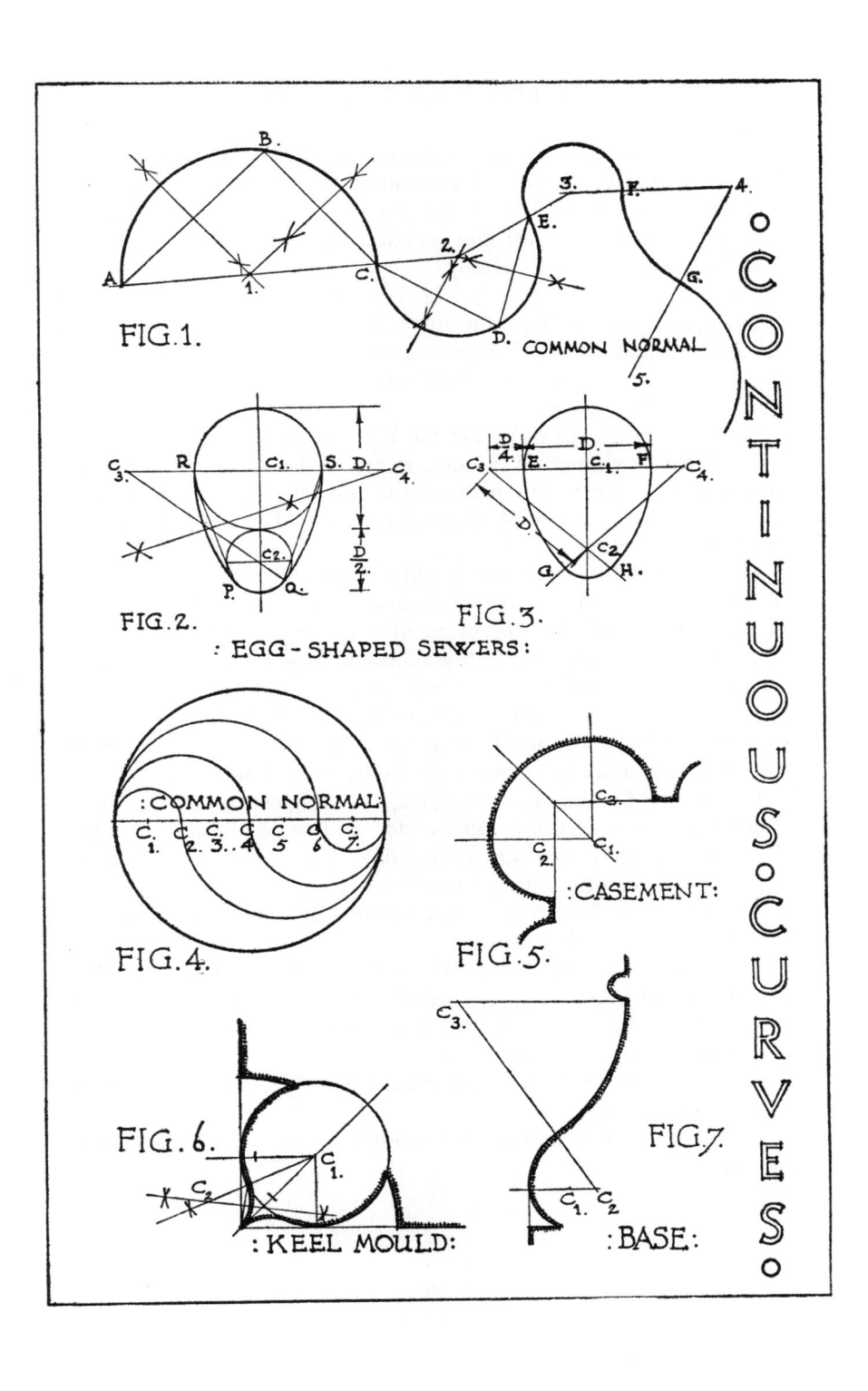

CONTINUOUS CURVES

FIG. 1.

FIG. 2.
FIG. 3.
: EGG-SHAPED SEWERS :

FIG. 4. : COMMON NORMAL :

FIG. 5. : CASEMENT :

FIG. 6. : KEEL MOULD :

FIG. 7. : BASE :

CONTINUOUS CURVES

A FURTHER selection of continuous curves as they apply to mouldings is given here. These are intended for practice and should be sketched to a larger scale. Choose a series of points in the curve and obtain a geometrical construction as near as possible to the sketched or designed moulding.

The bolection moulding in Fig. 1 was drawn in this manner and the construction of the main curve found by choosing the points A, B, C, D, and E. B–1 is a common normal to curves AB and BC. C–2 is a common normal to BC and CD and 2–3 is a common normal to CD and DE.

The remaining examples are Gothic mouldings.

In Gothic architecture mouldings are of the greatest importance and constitute one of the main keys to the style, sometimes, in fact, being the only means of determining the date of a mediaeval building.

Classic mouldings in stone usually project beyond the face of the building and produce strong contrasts of light and shade. In Gothic work, with the exception of drips and string courses, the stone mouldings are invariably recessed in grouped orders, producing shadows buried in deep recesses.

Figs. 2 and 3 are examples of hood and drip moulds to doors and windows and are also used for string courses to divide lofty walls into stages, as may be seen on the exterior of church towers.

Fig. 2, the first of these sections, is a thirteenth-century Early English example of the scroll mould, said to resemble a scroll of thick paper with overlapping outer edge. Fig. 3, the other example, is in the fourteenth-century Decorated style.

The curve in Fig. 5 is a compound ogee or wave moulding, typical of Decorated work.

Gothic capitals were usually of two kinds: moulded and floriated. Fig. 8 is an example of the former in the fifteenth-century Perpendicular style. Fig. 6, also in the same style, shows a base to a shaft and consists of several members.

Fig. 7 shows a roll-and-fillet mould usually associated with the Decorated period.

Fig. 4 illustrates a moulded window jamb of the Decorated period.

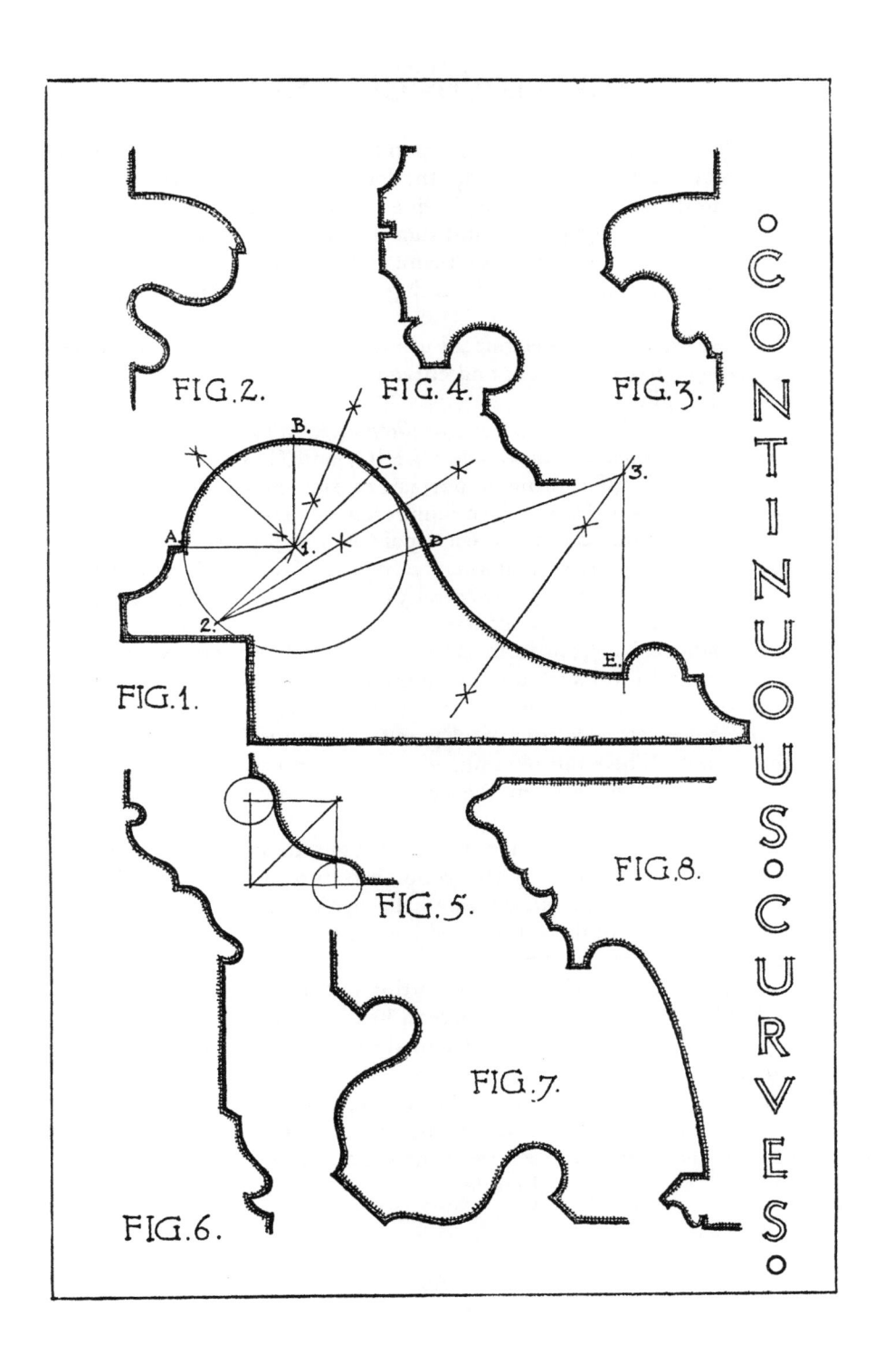

FIG. 2.

FIG. 4.

FIG. 3.

FIG. 1.

FIG. 5.

FIG. 8.

FIG. 6.

FIG. 7.

CONTINUOUS CURVES

LOCI: LOCUS OF CENTRES

A LOCUS is the path traced by a point moving according to certain fixed conditions. For instance, the locus of a point travelling at a constant distance from a straight line is a line parallel to the original line. The locus of a point equidistant from two intersecting straight lines bisects the angle between them. The locus of a point revolving about another point and maintaining a constant distance from it is a circle.

These are simple cases, but the examples illustrated involve points travelling between straight and curved lines, also between curves of unequal radii.

To trace the locus of a point travelling between two circles and equidistant from their circumferences: Let C_1 and C_2 be the two given circles, Fig. 1. Measuring on normals to each circle, set out a series of equal divisions from their circumferences. Draw arcs from these points; the intersection of each pair is a point in the required locus. It will be seen that the arcs marked 1 do not intersect but obviously the locus will pass centrally between the circles as it crosses the line connecting their centres.

A spandril has been formed in Fig. 2 by a horizontal, a vertical, and a curved line; such a spandril often occurs above arches. It is required to find the intersection line between the mouldings.

Set out a series of equal and parallel divisions from each of the three lines. Their intersections give points in the required mitres. The intersection between the two straight lines at 90° will be a 45° mitre.

The intersection of curved and straight moulded sections, of equal widths, obtained by the same procedure is shown in Fig. 3; it will always be curved, provided the sections are of equal width. Two curved mouldings also produce a curved intersection, unless the curves are of equal radii.

One application of loci to the setting out of tracery is illustrated in Fig. 4. There are two equilateral headed lights under one main equilateral arch. It is required to draw a circle tangential to all three arches.

Set out equal divisions parallel to the two curves to which the circle is to be tangential, using enough of them to give a locus of centres that will intersect the centre line of the arch giving the centre C of the required circle.

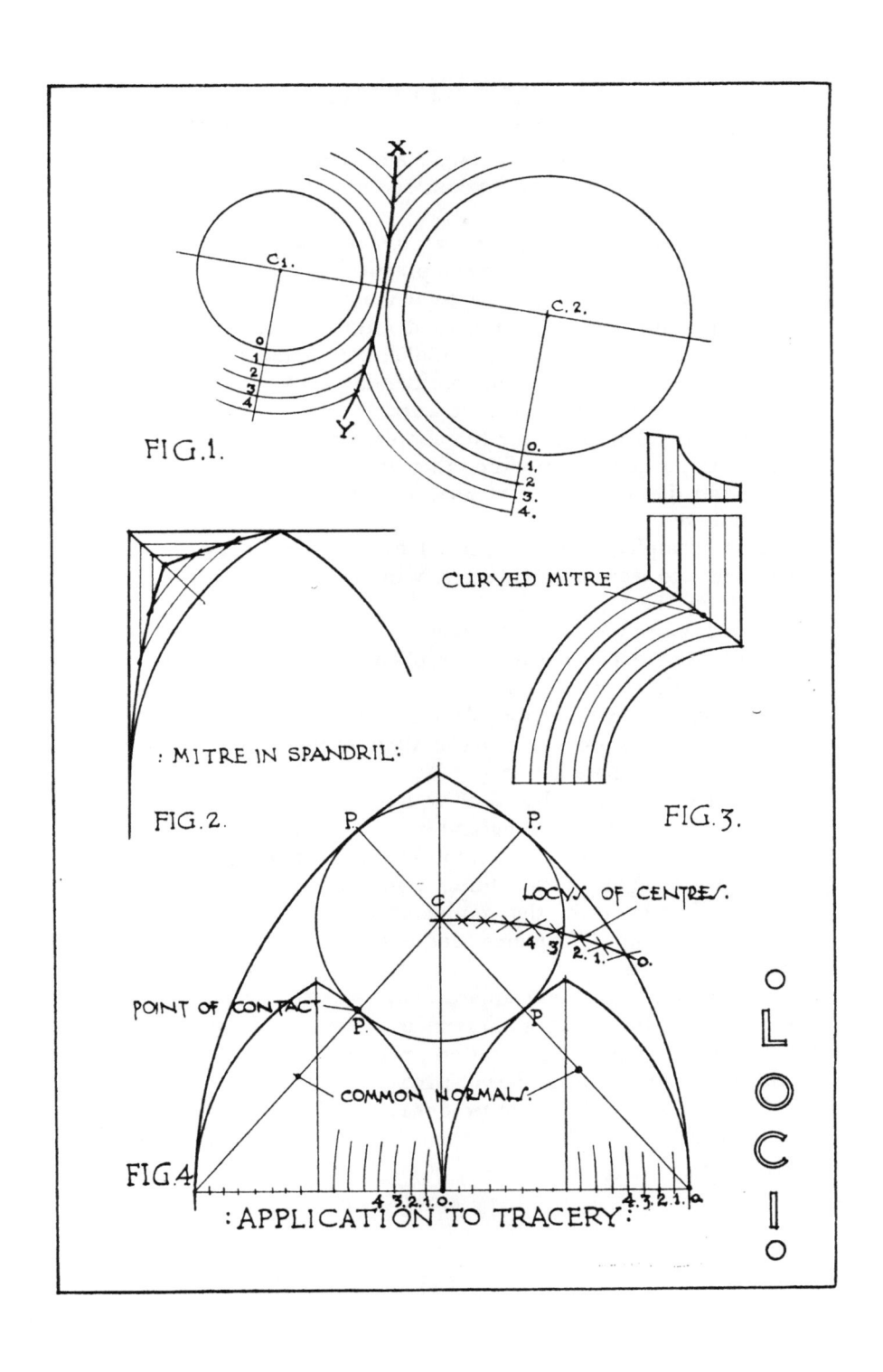

FIG. 1.

CURVED MITRE

MITRE IN SPANDRIL

FIG. 2.

FIG. 3.

LOCUS OF CENTRES.

POINT OF CONTACT

COMMON NORMALS.

FIG. 4.

: APPLICATION TO TRACERY :

LOCI

FRET PATTERNS

MANY patterns involve the circle and tangent constructions just dealt with; it is convenient to consider also at this stage those based on frets and squares.

The construction of patterns geometrically is a useful aid to draughtsmanship apart from the practical value.

These drawings should be copied not only with the object of producing the pattern but as a definite drawing exercise. The accurate drawing of a series of parallel, horizontal, and vertical lines of equal tone calls for quite considerable skill. This exercise also lends itself to brush and colour work.

There is justification for the simpler examples being attempted earlier in the course, providing they are reproduced to a fairly large scale and the importance of line, tone, and accurate measurement remembered.

The examples shown in Figs. 1 to 5 are known as FRET or KEY patterns and were used extensively in Greek architecture and previously by the Egyptians.

Based entirely on straight lines this form of ornamentation was usually executed in marble or applied colour but is by no means limited to these two media.

The bands may be on one plane surface, the effect being obtained by contrasting colours or inlays of different materials, or by forming sinkings of alternate bands.

The patterns being based on squares are also particularly adaptable to floor and wall tiling and mosaic cubes.

The drawing should be commenced by making a complete grid by means of lightly drawn lines, lining in the actual pattern afterwards. The width of the completed pattern is indicated in each case as a multiple of squares, and examples 4, 5, 7, and 9 squares wide are shown.

For convenience these squares may very well be made $\frac{1}{8}$ inch or $\frac{1}{4}$ inch wide but as an additional problem an odd width may be taken as the total width of pattern, necessitating the division of the width into an equal number of parts.

The use of graph paper produces much quicker results but is not advocated as it eliminates the value of the exercise in draughtsmanship.

FIG. 1.

FIG. 2.

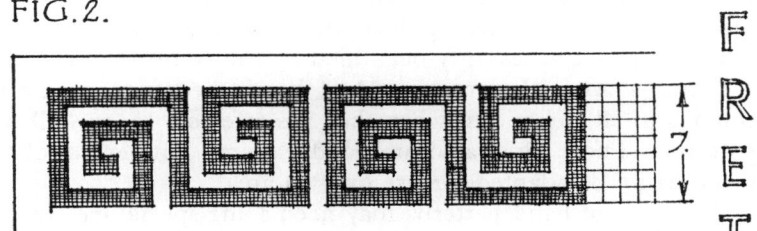

FIG. 3.

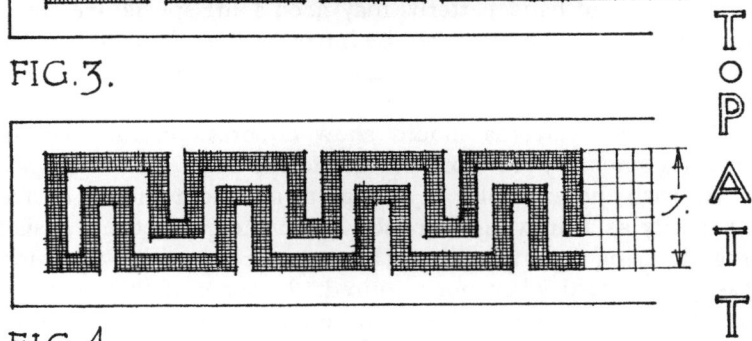

FIG. 4.

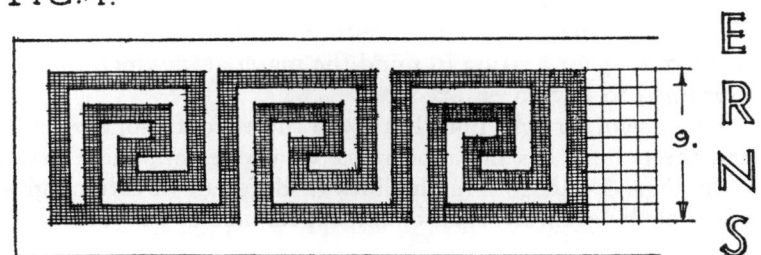

FIG. 5.

OF FRET PATTERNS

PATTERNS BASED ON SQUARES

THE number of patterns based on straight lines that could be devised is limitless, particularly when colour is used to add to or emphasize the effect.

Those illustrated here, for instance, are all based on the one geometrical system given in Fig. 1, namely, a square diagonally within another square.

Patterns, as with ornament generally, should be used with discretion, and whilst their design or reproduction in geometry is valuable their use in practice should be limited and undertaken only by a person with experience in design.

Consideration must be given to position, surroundings, and materials of construction as well as the actual pattern. A little ornamentation correctly applied beautifies, but a thing that is not beautiful is not necessarily beautified by the addition of ornament.

Patterns are either composed of bands or strips or are continuous in all directions, when they are referred to as "all-over" patterns. Those illustrated are band patterns but several of them could easily be adapted for use over an area.

The ends of band patterns may need some special treatment, as in Fig. 8, and a variation can be introduced at the centre. Corners also usually present certain difficulties necessitating variations in the pattern.

"All-over" patterns should show no break in their continuity but run smoothly from one feature to the next.

As with the examples of fret patterns, these should be treated primarily as a drawing exercise. Form the geometrical basis first, Fig. 1, using very fine lines, the unwanted portions of which can easily be erased when not required. In some of the figures a few additional construction lines will be necessary and are obtained by subdividing the original squares.

The effect may be obtained either by some form of shading or by colour, again bearing in mind the material in which the work is presumed to have been executed.

The use of straight lines has given the effect of solidity in Fig. 7 and the pattern has the appearance of a series of square prisms.

In Figs. 5 and 10 straight lines in different directions give the effect of the grain in wood or the appearance of parquetry.

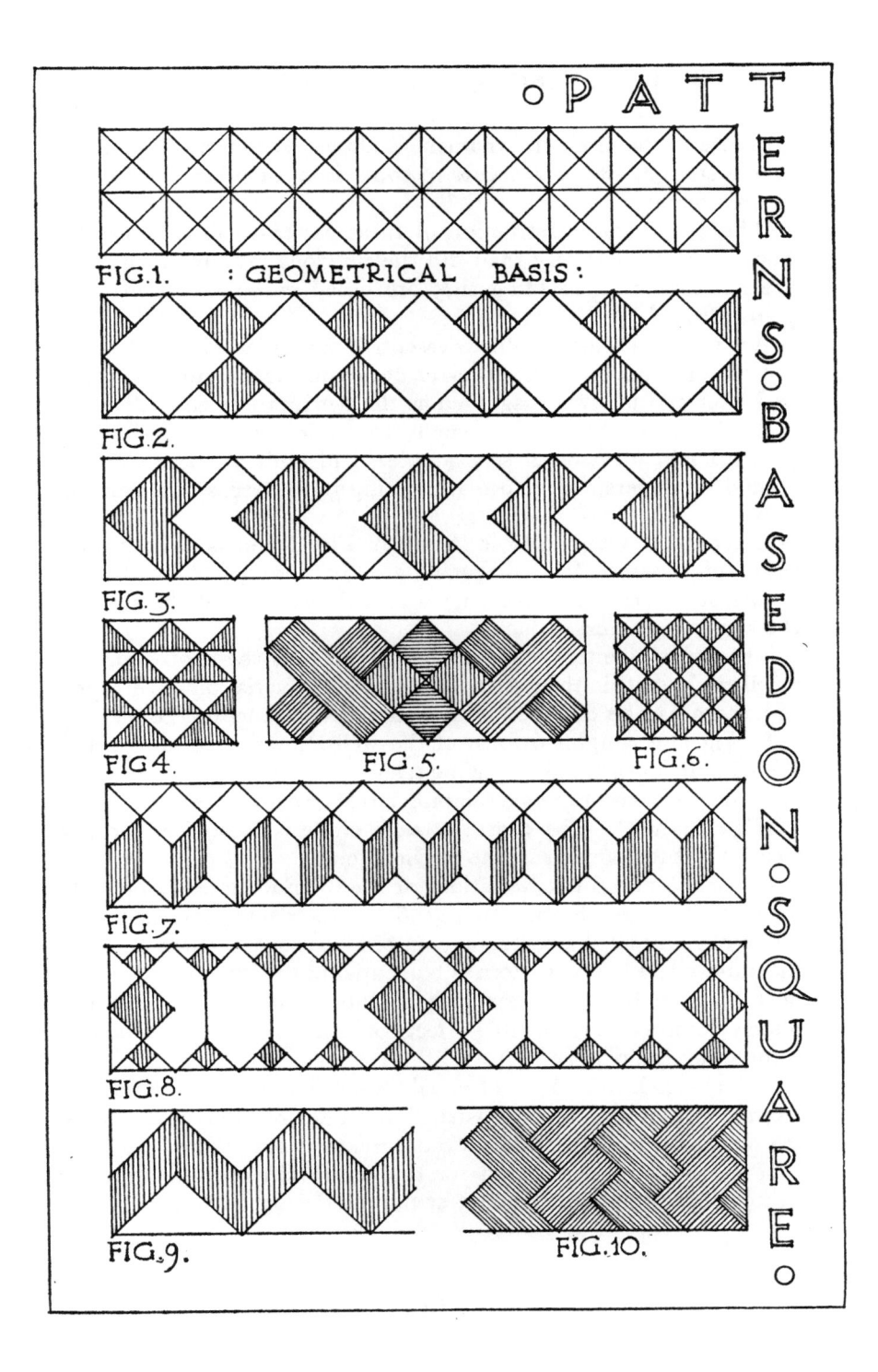

PATTERNS BASED ON SQUARE

FIG.1. : GEOMETRICAL BASIS :

FIG.2.

FIG.3.

FIG.4.   **FIG.5.**   **FIG.6.**

FIG.7.

FIG.8.

FIG.9.   **FIG.10.**

PATTERNS BASED ON CIRCLES

THE use of the circle in ornament is almost as common as the use of the straight line and occurs in some form or other in all styles of architecture.

The design and construction of these patterns call for a greater knowledge of geometry than did the previous examples, but they are not difficult, providing they are based on sound geometrical principles.

Accurate draughtsmanship is essential, particularly in the setting-out of construction lines; an error occurring here cannot be rectified later and is often magnified as the work proceeds.

Circles in contact and tangents to circles are the underlying principles in most of these examples, and both of these subjects should be understood before attempting the patterns as a drawing only.

A type of wave pattern is shown in Fig. 1 and is set out on a series of squares. The diagonals are the common normals and points of separation of the curves and the corners of the squares the striking centres.

The point of intersection of the diagonals is the centre for the small circles within the pattern. Considerable variation of this type of pattern can be obtained with a similar constructional basis.

In Fig. 2 a simplified form of the well-known bead and reel is shown and consists only of semicircles.

Two variations of the overlapping circle or coin pattern are given in Figs. 3 and 4. The centre of each circle is located at the extremity of the diameter of the previous one.

In Figs. 5 and 6 two variations of the interlacing circle or guilloche pattern are shown. In Fig. 5 the interlacing circles are of different radii; all the centres are on one horizontal centre line and the outer circle in one pattern is tangential to the inner circle in the next. It is advisable to draw both the inner and outer circles completely in faint line to ensure perfect continuity of the curves before completing the pattern.

The circles in Fig. 6 are of equal radii otherwise there is no difference in the method of construction. The additional band lines are tangential to each other on the horizontal centre line.

If some filling is required for the centre circles these may be treated separately, independent of the interlacing circles.

PATTERNS BASED ON CIRCLE

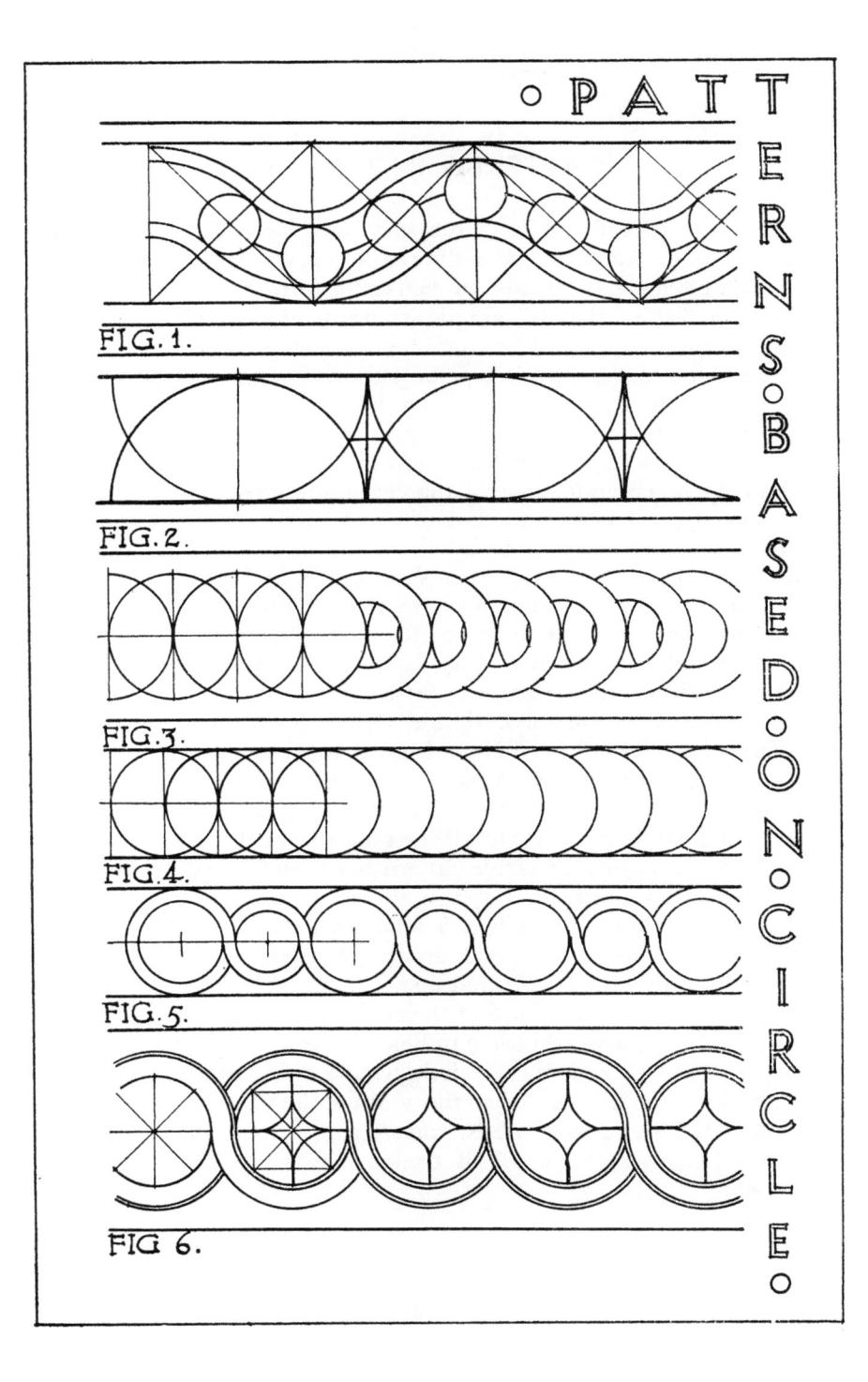

FIG. 1.

FIG. 2.

FIG. 3.

FIG. 4.

FIG. 5.

FIG 6.

PATTERNS IN CIRCLES

PATTERNS enclosed within a circle may form the centre feature of a design or be a feature in themselves as in a tiled floor. Again, a series of such circles may be used in contact to form a band pattern, or spaced and connected by a simple arrangement of straight or curved lines to form an all-over or continuous pattern.

A small selection of this type of pattern is given in Figs. 2, 4, 6, and 8 with the basis of construction in each case. As with previous examples, they form a basis for a great number of variations.

Two interlacing triangles are used in Fig. 2, which is set out by dividing the circumference of the circle into six equal parts, each divisional point being the apex of one triangle. The thickness of the band is governed by the intersection of the side of one triangle with the median of the other as shown at X in Fig. 1.

A similar basis has been used in Figs. 3 and 4 as the primary construction, a semicircle being afterwards drawn on each side of each triangle, as shown in Fig. 3. The band thickness is drawn within each of the original semicircles.

The setting out of the design in Fig. 6 should be obvious after reference to Fig. 5. Each circle has a diameter equal to the radius of the large circle which gives a common intersection point at the centre. The circles are all tangential to the large circle but not tangential to each other. It will be seen that without the addition of the band a quite attractive pattern has been formed by the construction lines alone.

The pattern shown in Fig. 8 relies entirely on the line for its effect and makes an excellent drawing exercise particularly when carried out in various coloured inks.

Divide the circumference of the circle into any number of equal parts—twelve have been taken in Fig. 7. Decide on the width of the band, as at BD, and draw a circle having a diameter equal to the diameter of the first circle less the width of the band, as at centre C. This is the first circle, and by transferring point C to each of the other diameters the series of twelve interlacing circles may be drawn.

PATTERNS IN CIRCLES

FIG. 1.

FIG. 2.

FIG. 3.

FIG. 4.

FIG. 5.

FIG. 6.

FIG. 7.

FIG. 8.

THE ELLIPSE AS A PLANE FIGURE

The ellipse, as a plane figure, is defined as the locus of a point moving so that the *sum* of its distances from two fixed points, called the focal points, is constant. This sum is equal to the length of the major axis of the ellipse. An ellipse is also the outline of the section revealed by an inclined cutting through a cone.

A semi-ellipse is shown in Fig. 1. The focal points are F and F_1. The MAJOR AXIS (AB) and MINOR AXIS (twice CD) may bear any proportion to each other, but are always unequal.

To find the focal points of a given ellipse, swing arcs from one extremity of the minor axis (D in the figure) using as radius one-half the major axis.

DF and DF_1 are the focal distances, and the angle FDF_1 is the focal angle, for point D. A NORMAL to an ellipse at any point (such as G in the figure) bisects the focal angle, and makes a right angle to the tangent at that point.

Applying the definition of an ellipse to Fig. 1 we see that FD + DF_1 = FG + GF_1 = FB + BF_1 = the major axis AB. This provides a principle for constructing the curve.

For a large ellipse set out major and minor axes, fasten a piece of string to nails driven in at the focal points, making the length of the string equal to the major axis. By holding a pencil in the angle formed by the string stretched taut, and moving it round, the curve will be traced.

For small-scale drawing-board work, requiring greater accuracy, this principle may be adapted as in Fig. 3. Divide the distance between F and F_1 into any convenient number of parts. Using centres F and F_1 and radii measured A–1 and B–1 (together equal to AB) swing arcs, locating point C on the curve. Repeat with another pair of radii A–2 and B–2, also together equalling AB; and so on.

The curve in Fig. 1 has been drawn by the trammel method in the following manner: Take a strip of paper and mark on it half the major and half the minor axes, so that the distance C–C_1 equals the difference between the two half-axes. By revolving the trammel strip, keeping points C and C_1 in contact with the axes lines, the point AD will trace the curve.

A method of drawing an ellipse in a rectangle is shown in Fig. 2; the divisions on the major axis and on the side of the rectangle are equal in each group.

In Fig. 4 the two circles have radii equal to half the major and half the minor axes respectively. The points of intersection of the horizontal and vertical lines define the curve.

For the area of an ellipse, see page 108.

THE ELLIPSE AS A PLANE OF FIGURE

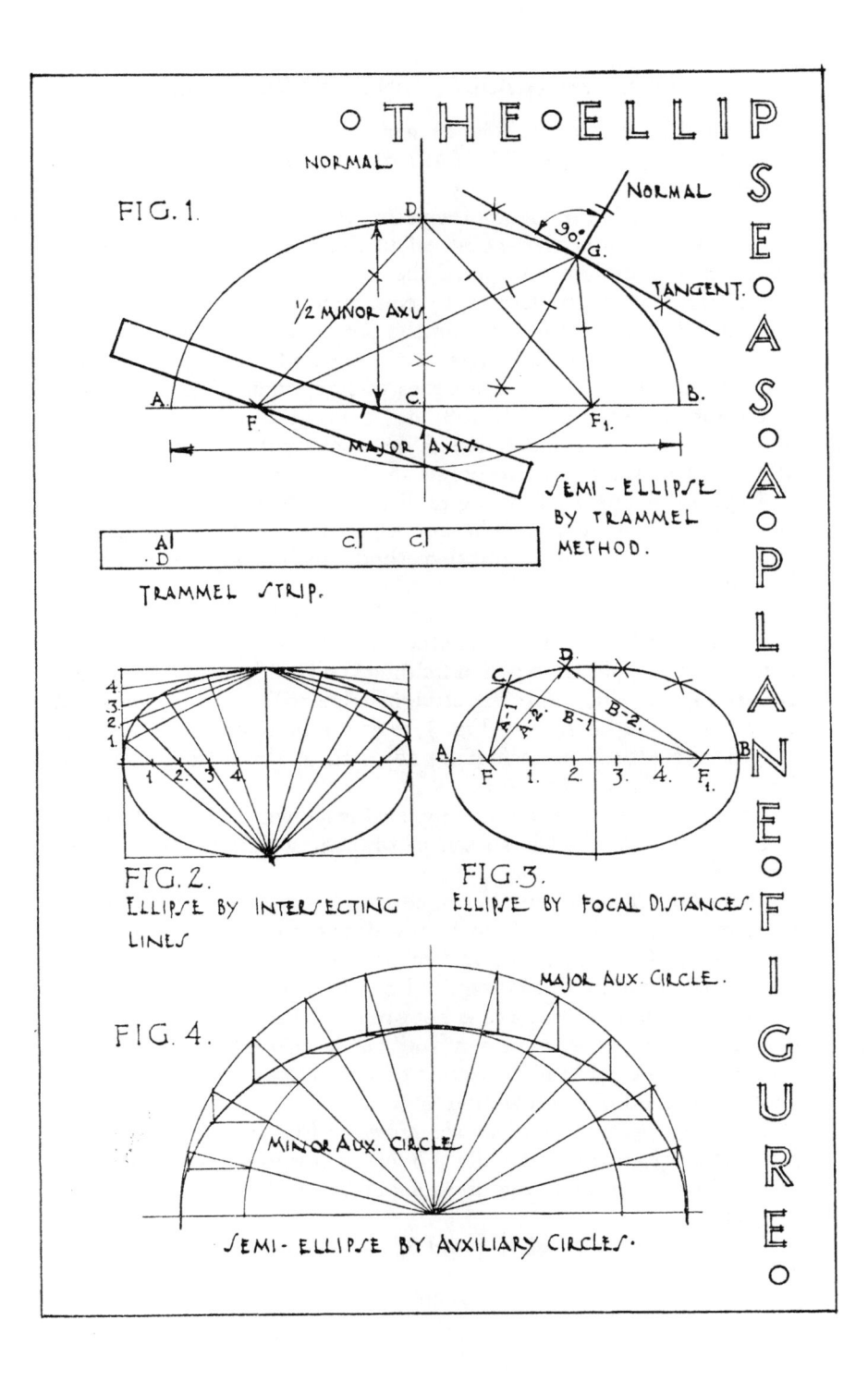

FIG. 1.

NORMAL — NORMAL — TANGENT
90° G
½ MINOR AXIS
A — F — C — F₁ — B
MAJOR AXIS
SEMI-ELLIPSE BY TRAMMEL METHOD.

TRAMMEL STRIP.
A / C / C
D

FIG. 2.
ELLIPSE BY INTERSECTING LINES

FIG. 3.
ELLIPSE BY FOCAL DISTANCES.
C D
A-1, A-2, B-1, B-2
A — F — 1 2 3 4 — F₁ — B

FIG. 4.
MAJOR AUX. CIRCLE.
MINOR AUX. CIRCLE
SEMI-ELLIPSE BY AUXILIARY CIRCLES.

THE PARABOLA AND HYPERBOLA

THE parabola and hyperbola, as sections of a cone, are dealt with at a later stage (pages 152, 154) and are considered here only as plane figures.

A PARABOLA is the locus of a point travelling in a plane so that its distance from a fixed point called the *focus* is equal to its distance from a straight line called the *directrix*.

In Fig. 1 the problem is to construct a parabola, given the directrix AB and the focus F on the perpendicular HK.

The first point in the curve is G, situated midway between F and H. Mark out on GK a series of horizontal divisions (which need not be equal) as at 1, 2, 3, etc.

With radius H–1 and centre F draw arcs cutting the horizontal division 1 in L and M two points in the curve.

Repeat this operation, using radii H–2, H–3, etc., and always the focal point F as centre. The curve is a freehand line connecting these points, but a mechanical method can be adopted, as shown in Fig. 3.

A straight edge is laid along the directrix AB and a square on the straight edge. If a cord is attached by a nail to the end of the square and to point F, a pencil holding the cord in contact with the square as it is slid down the straight edge will draw the required curve. Point G, as seen in Fig. 1, is at the bisector of HF and the length of the cord must first be adjusted at this point, its actual length being X–G plus G–F.

Another position of the moving cord is shown in the angle FPX.

In Fig. 2 a method is shown of drawing a parabola in a given rectangle ABCD.

Divide DE into any number of equal parts and DA into the same number of equal parts. Radial lines drawn from the divisions on AD to point G intersect the vertical lines from the divisions on DE at a series of points in the required parabolic curve.

The HYPERBOLA as a curve is not often used in building, and only one method of constructing it will be described here. It is the outline of a special kind of section of a cone (see page 154, where another method of construction is given).

To draw a hyperbola, given the diameter EC, Fig. 4, *an ordinate* AB *and an abscissa* CB: Divide AB and AD into the same number of equal parts. Radiate the divisions on AB to E and the divisions on AD to C.

The radial lines 1 and 1, 2 and 2, etc., intersect at points in the required curve.

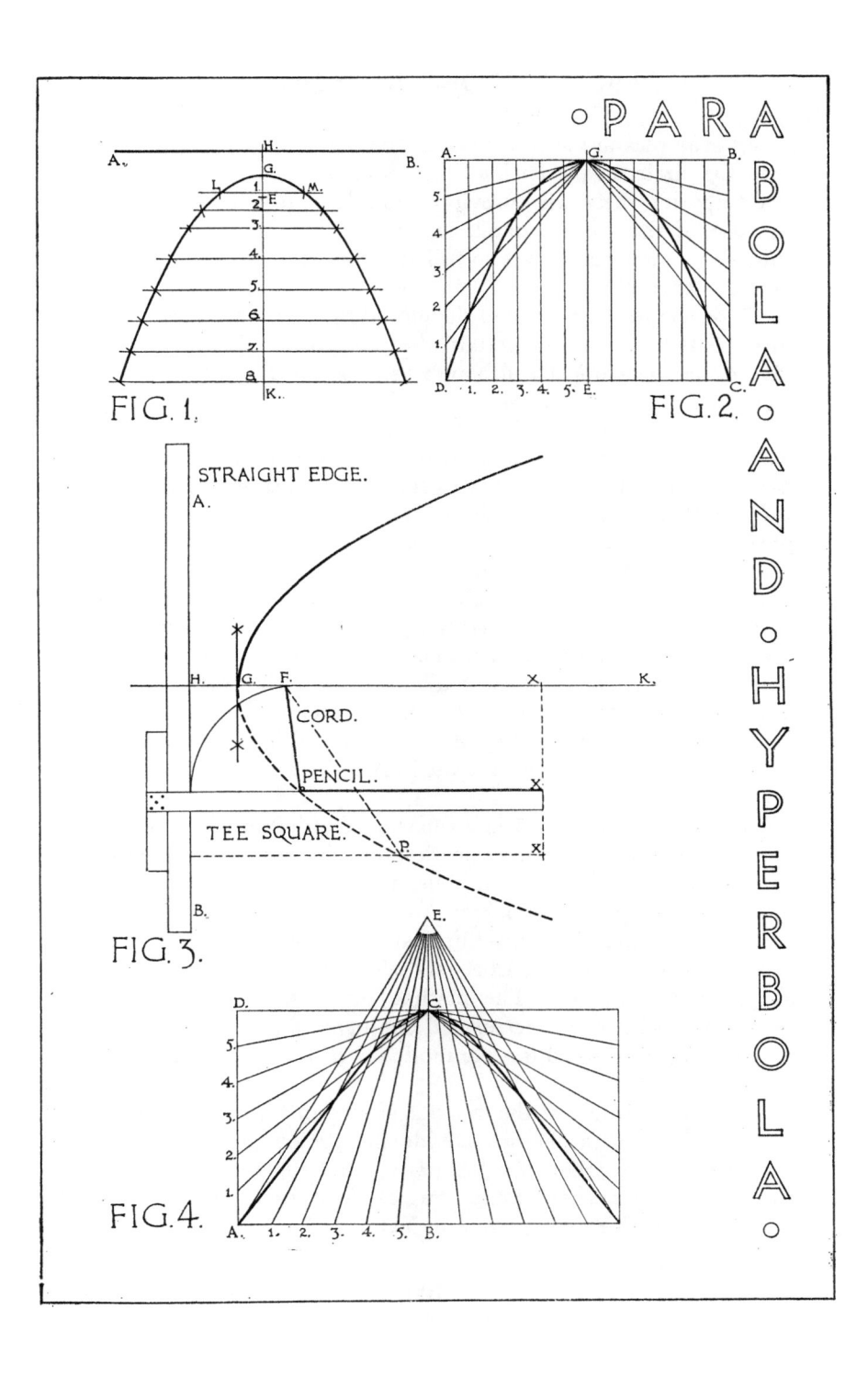

PARABOLA AND HYPERBOLA

FIG. 1.

FIG. 2.

FIG. 3.
STRAIGHT EDGE.
CORD.
PENCIL.
TEE SQUARE.

FIG. 4.

CONIC SECTIONS IN MOULDINGS

SOME examples are shown here of the use of conic sections, other than the circle, in the design of mouldings.

Comparing these with previous mouldings it will be noticed that the use of the ellipse, parabola, and hyperbola produces a considerably flatter curve, not so full and robust as those composed of portions of circles.

These shapes are typical of the mouldings used in Greek architecture. It is by no means certain that the Greeks did actually draw the curves mechanically although there is no doubt they were inspired by the conic sections.

The distance a moulding stands forward horizontally from the wall or adjacent member is termed its projection. This term should not be confused with the usual geometrical meaning of the word.

Each of the sections shown in Figs. 1–6 consists of one or more portions of curves already dealt with; their construction therefore should need little explanation.

A CYMA RECTA in a rectangle is given in Fig. 1. Bisect the sides of the rectangle in both directions and divide each bisector into the same number of equal parts. Vertical lines drawn from the horizontal divisions intersect radial lines from the vertical divisions at points in the two curves, which are portions of a parabola.

The CYMA REVERSA in Fig. 2 is constructed in the same manner, but the curve being in reverse the horizontal and radial lines are in reverse positions.

The CAVETTO in Fig. 3 is a quarter of an ellipse in a rectangle. Divide the height and the projection into the same number of equal parts. Radiate the projection divisions from point Y and the height divisions from point X, making X4 equal to Y4.

The projection of a semi-elliptical SCOTIA is shown in Fig. 4 and consists of a semi-ellipse in a parallelogram and not in a rectangle as previous examples. The method of construction remains the same. The diameters are termed conjugate diameters.

The OVOLO in Fig. 5 is a parabolic curve similar to half the cyma reversa.

A second example of the ovolo is shown in Fig. 6 and has a projection equal to its height. Divide the projection and the height into the same number of equal parts. Radiate the projection divisions from the corner of the square, Y, and the height divisions from point X, XY being equal to YO.

CONIC SECTIONS IN MOULDINGS

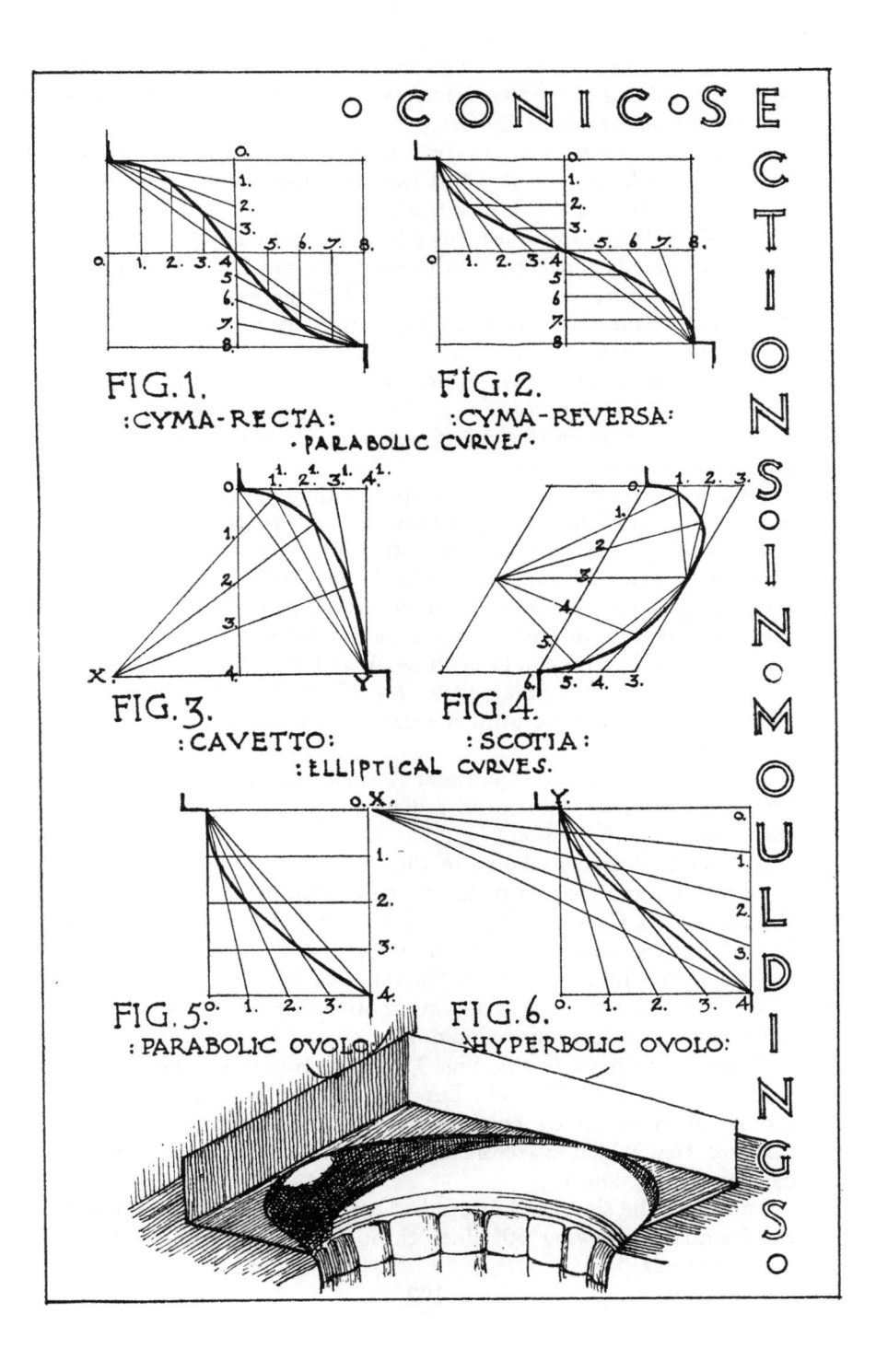

FIG. 1. :CYMA-RECTA:

FIG. 2. :CYMA-REVERSA:

:PARABOLIC CURVES:

FIG. 3. :CAVETTO:

FIG. 4. :SCOTIA:

:ELLIPTICAL CURVES:

FIG. 5. :PARABOLIC OVOLO:

FIG. 6. :HYPERBOLIC OVOLO:

APPROXIMATIONS TO ELLIPSE

In a true ellipse no portion of the curve is a portion of a circle.

If a line is drawn parallel to an ellipse the curve is not also an ellipse.

These two facts raise difficulties in practical setting-out and make the use of a curve struck from several centres or consisting of several arcs of circles desirable.

In arches, particularly in brickwork, setting-out the normal joints to an ellipse is a laborious task, and the fact that the intradosial and extradosial (i.e. inside and outside) curves are not both elliptical is also likely to lead to mistakes, particularly when more than one trade is concerned with the setting-out.

If, for example, there is a curved head to the wood door or window-frame beneath a stone arch, of the several parallel curves which is to be the true ellipse?

Usually the architect would decide on the soffit of the arch, but unless the men responsible for the setting-out of their various materials are clear on this point and work from a common agreed curve mistakes may occur.

The examples shown give fair approximations to a true ellipse, but a greater number of points may be found to give even greater accuracy.

In Fig. 1 five centres are used after first locating points in the curve as previously explained, Fig. 2, page 97. After enclosing the span and the rise in a rectangle divide half the span and the rise into the same number of equal parts. By means of radial lines obtain a number of points such as A and B. Bisect DB, which bisector gives C_1, the first centre. Join C_1B as a common normal. Bisect AB and produce until it meets the common normal in C_2, the second centre.

Join C_2A, the second common normal, and bisect AE to find centre C_3. Normal joints will radiate to the respective centres of the portions of the curve.

A second method is shown in Fig. 2.

Obtain C_1 and C_1B as in the previous example, also C_2. Draw the horizontal C_2G, cutting the circular arc BA produced in G. Join GE and produce until it cuts the arc in F. Join FC_2, cutting the major axis in C_3 the centre for the arc FE.

The examples, Figs. 3 and 4, are more easily constructed but do not give such a near approximation.

Draw the major axis AB, Fig. 3, and the minor axis DE.

Draw a circle on AB, join DA and DB and with centre D and radius DF draw an arc cutting DA and DB at H and G. Bisect GB and HA, which bisectors are the common normals and give centres C_1, C_2, and C_3.

In Fig. 4, the simplest but crudest method, the common normals are found by drawing 60° lines through the mid points of half the major axis.

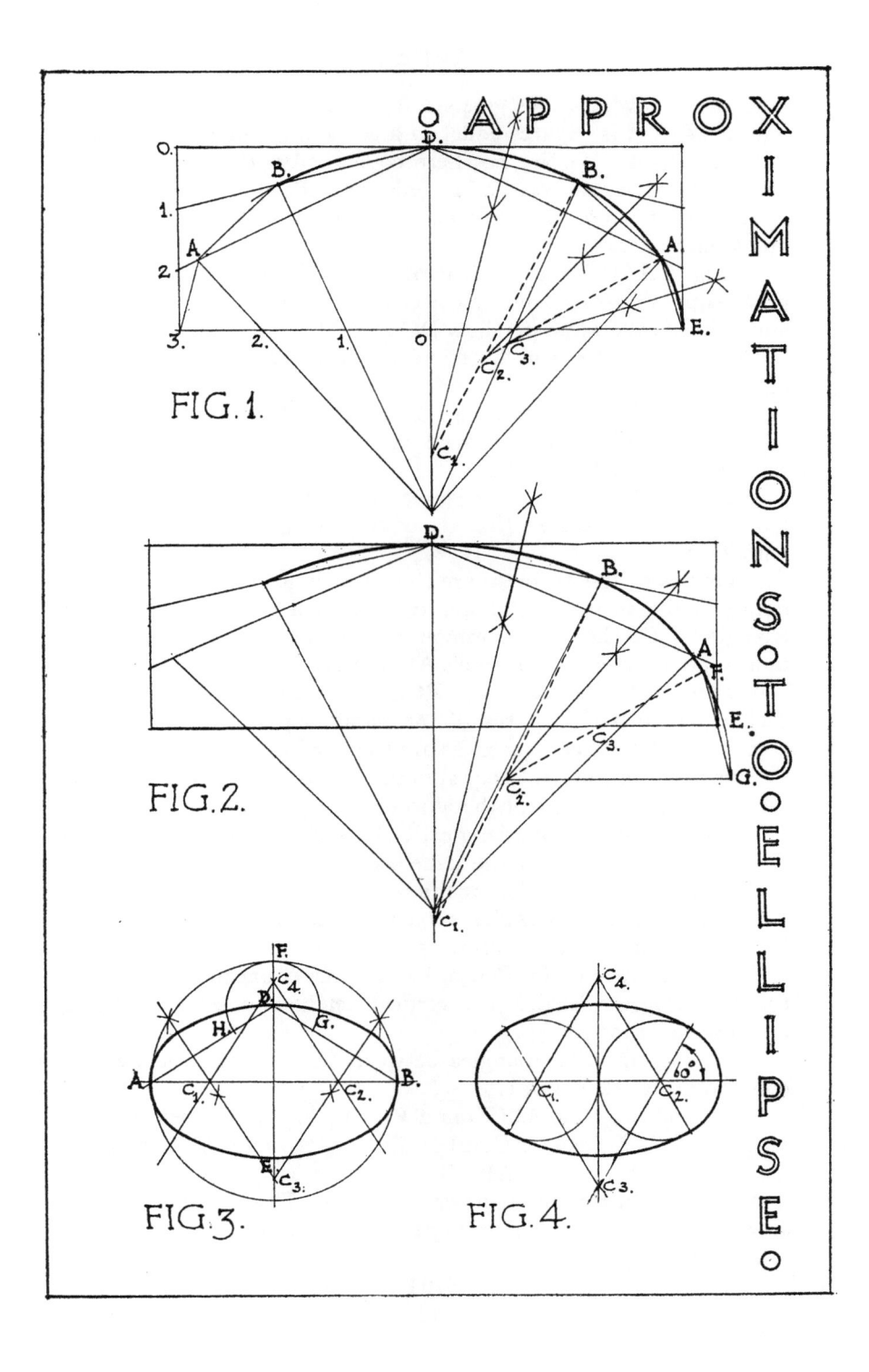

APPROXIMATIONS TO ELLIPSE

FIG. 1.

FIG. 2.

FIG. 3.

FIG. 4.

AREAS

THE two following pages give some graphical examples of equivalent areas, that is, figures equal in area to other figures though of different shape. We are not here concerned with calculating the magnitude of areas in units, but only with their *relative* sizes.

Parallelograms on the same base and between the same parallels are equal in area.

In Fig. 1 ABCD is a parallelogram, EFCD is another parallelogram equal in area to the first because it is on the same base DC and between the same parallels DC and AF.

Triangles on the same base and between the same parallels are equal in area.

ABC is a given triangle, Fig. 2. Triangle DBC is a triangle of equal area, it being on the same common base CB and between the same parallels CB and AD.

If a parallelogram and a triangle are on the same base and between the same parallels the area of the triangle is half that of the parallelogram.

Examples of this theorem are shown in Figs. 3 and 4. The first figure is the rectangle ABCD, the diagonal of which gives the triangle BCD. DC is a common base to both figures, and both figures are between the parallels AB and DC. The second example shows two triangles ECD and BCD which are equal in area and each equal to half the area of the parallelogram ABCD. DC is common to the three figures, as are the parallels AE and DC.

The area of a triangle is equal to the area of a rectangle upon the same base but having half the altitude.

In Fig. 5 the area of triangle ECD equals that of the rectangle ABCD because they have the same base DC and the vertical height of the triangle is twice that of the rectangle.

Parallelograms or triangles upon the same base have their areas in the same ratio as their altitudes.

Parallelogram ABCD, Fig. 6, has an area equal to twice that of the parallelogram EFCD, the vertical height of the former being 2X and the latter X.

Parallelograms and triangles between the same parallels are to one another as their bases.

The parallelograms ABCD and FGCE, Fig. 7, are between the same parallels DC and AG. Base EC is $\frac{1}{3}$ the base DC, therefore the area FGCE is $\frac{1}{3}$ area ABCD.

A second illustration is shown in Fig. 8. Triangle FBD has an area equal to half that of triangle ABC, their bases being in the ratio 1:2.

FIG.1. FIG.2. FIG.3. FIG.4. FIG.5. FIG.6. FIG.7. FIG.8.

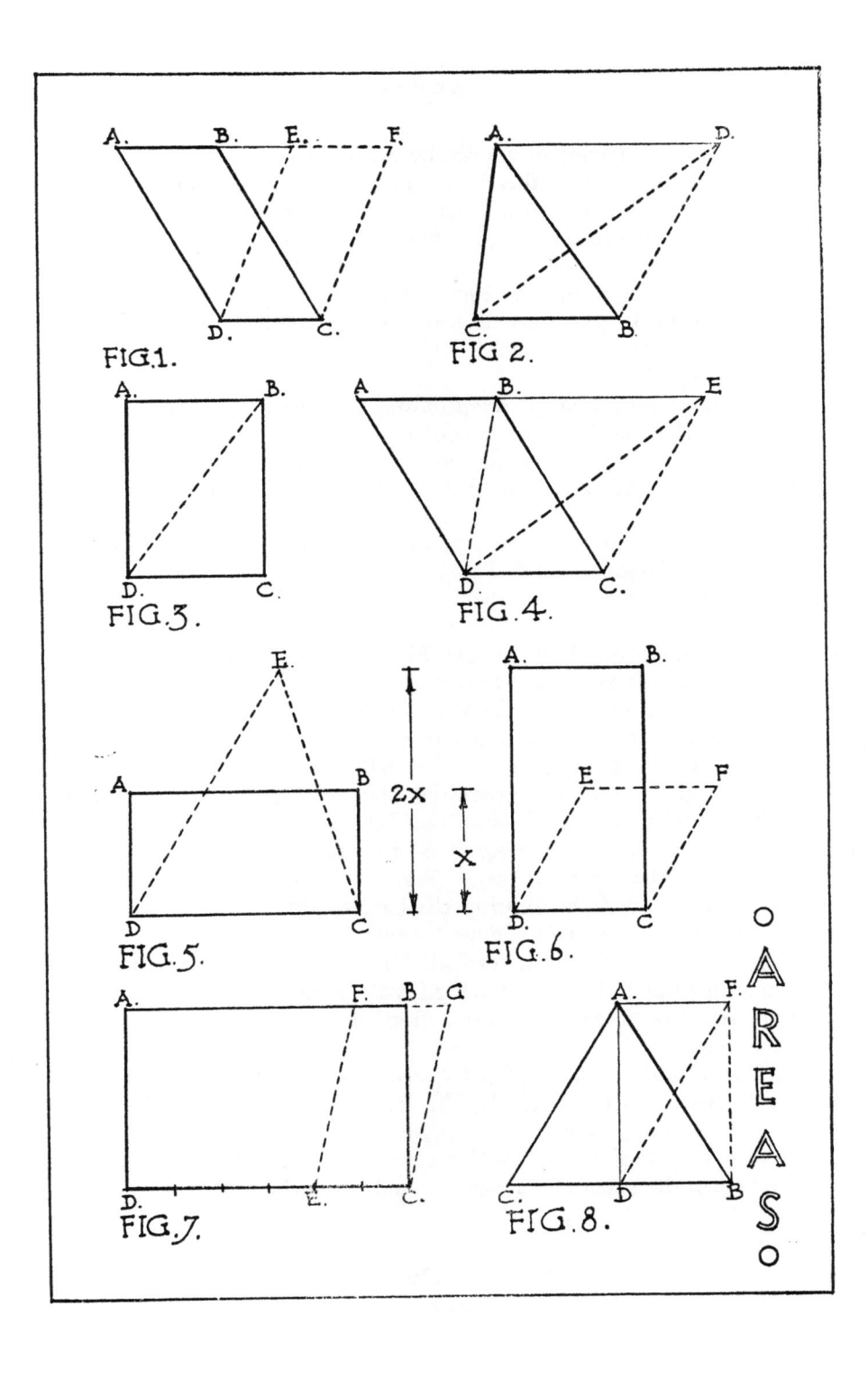

AREAS

AREAS

THE areas of irregular figures bounded by straight lines can be obtained by dividing the figure into a number of simpler forms such as triangles and rectangles. Graphical solutions are also possible by constructing simpler figures of equivalent areas, which may then be computed.

To construct a triangle equal in area to a given quadrilateral, Fig. 1: Join AC and from B draw BE, making it parallel to AC.

Join AE, when triangle AED is equal in area to the given quadrilateral ABCD.

To construct a triangle equal in area to a given irregular polygon: Let ABCDE be the given irregular polygon, Fig. 2. This problem requires the use of the previous method on either side of the figure. Join AC and from B draw BG, making it parallel to AC, join AG.

Repeat this operation with points ADE and F, when AFG is the required triangle.

To construct a triangle equal in area to a given triangle and of a given altitude: Let ABC be the given triangle and DG the given altitude of the required triangle, Fig. 3. Join D to B and C. Draw AE and AF, making these lines parallel to DB and DC respectively. Join DF and DE, when DEF is the required triangle.

To construct a triangle equal in area to a given triangle but on a given base of a different length: Let ABC be the given triangle and CD the given base of the required triangle, Fig. 4. Join DA, draw BE parallel to DA, and join ED. EDC is the required triangle.

To construct a square equal in area to a given rectangle: Let ABCD be the given rectangle, Fig. 5. With centre B and radius AB draw a quadrant, cutting the base of the rectangle produced at H. Bisect CH at O and with centre O and radius OH draw a semicircle. Produce the vertical side BA of the rectangle until it cuts the semicircle in E, which gives BE as one side of the required square. Complete the construction of the square in the usual manner.

To construct a square of given area without calculating the lengths of the sides: In the example, Fig. 6, suppose the required area is 3 square units. Draw a rectangle choosing sides of convenient length to give this area; in this case 3 units × 1 unit. Proceed to construct the square from the rectangle as in the previous example.

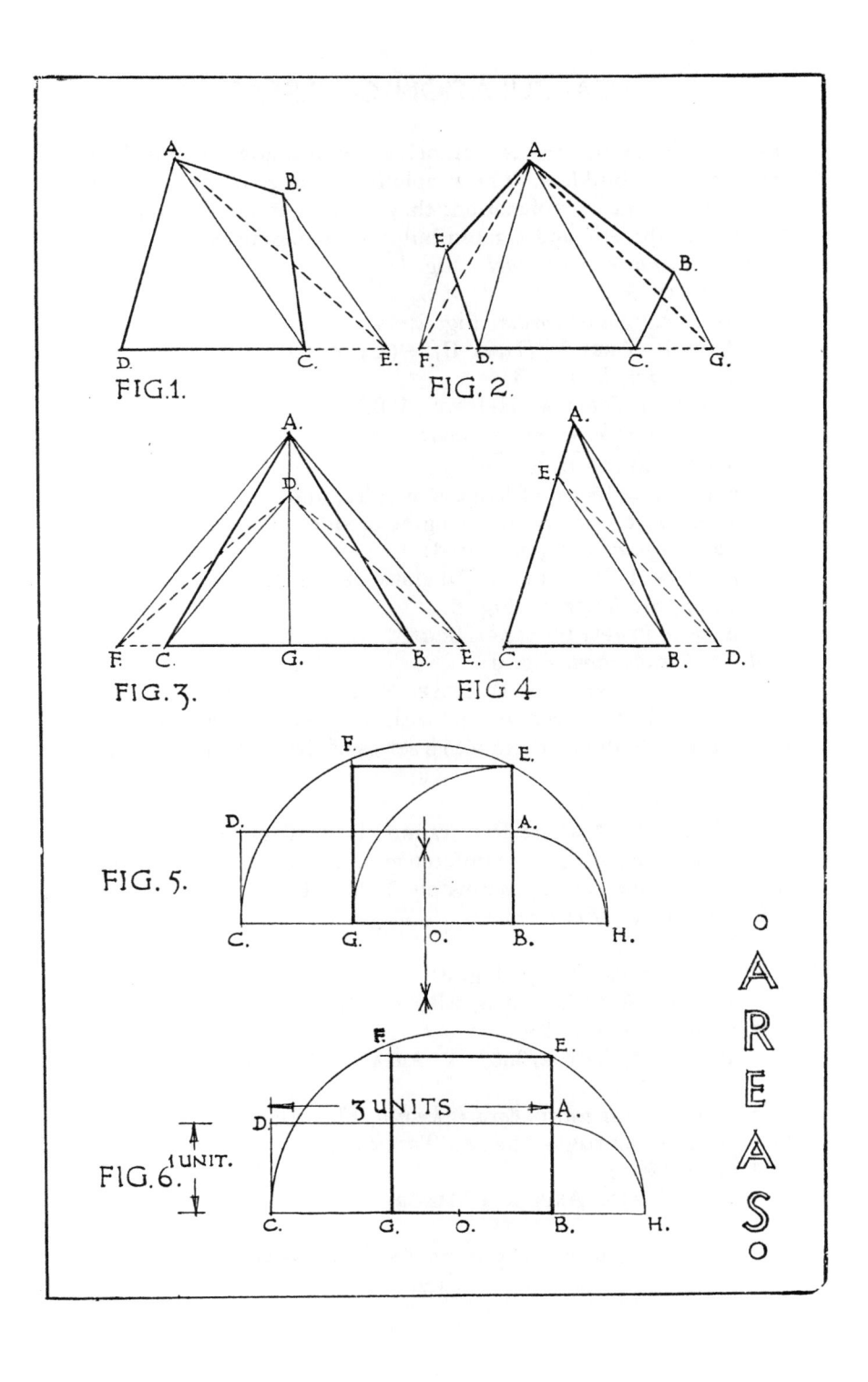

FIG. 1.

FIG. 2.

FIG. 3.

FIG 4

FIG. 5.

FIG. 6.   1 UNIT.   3 UNITS

AREAS

CALCULATION OF AREAS

THE calculation of areas is a branch of mathematics of considerable importance to builders. The simple formulae given here for reference demand little explanation: they assume an elementary knowledge of arithmetic and algebra but not trigonometry.

Area of square or rectangle, Fig. 1:
  LENGTH × BREADTH. L × B.
Area of margin or border, Fig. 2:
  AREA 1 − AREA 2. $(L_1 \times B_1) - (L_2 \times B_2)$.
Area of triangle, Fig. 3:
  $\frac{1}{2}$ BASE × VERTICAL HEIGHT. $\frac{1}{2}$ B.H.
Area of triangle, second method:
  $\sqrt{s(s-a)(s-b)(s-c)}$
  where $s = \frac{1}{2}$ sum of lengths of sides, and
  a, b, and c = respective lengths of sides of triangle.
Area of regular polygon, Fig. 4:
  AREA OF 1 TRIANGLE × NUMBER OF SIDES.
Area of parallelogram, Fig. 5:
  BASE × PERPENDICULAR HEIGHT. B. × H.
Area of trapezium, Fig. 6:
  AVERAGE LENGTH OF PARALLEL SIDES × HEIGHT.

Note.—It is often convenient with irregular figures bounded by straight lines to divide them into a series of simple figures, triangles, rectangles, etc.

Area of circle, Fig. 7:
  πR^2 or $\pi D^2/4$ where R = radius, D = diameter.
  π is the ratio of the circumference of any circle to its diameter and may be taken as approximately $3\frac{1}{7}$ or $3 \cdot 142$.
Circumference of circle:
  $2\pi R$ or πD.
Area of annulus or ring, Fig. 8:
  AREA 1 − AREA 2, that is, $\pi R_1^2 - \pi R_2^2$.
Area of sector, Fig. 9:
  ANGLE OF SECTOR/360° × AREA OF CIRCLE, that is $x/360° \times \pi R^2$.
Area of segment when chord is known, Fig. 10:
  AREA OF SECTOR − AREA OF TRIANGLE.
Area of ellipse:
  $\pi \times \frac{1}{2}$ MAJOR AXIS × $\frac{1}{2}$ MINOR AXIS.

(For areas of curved surfaces, see page 110)

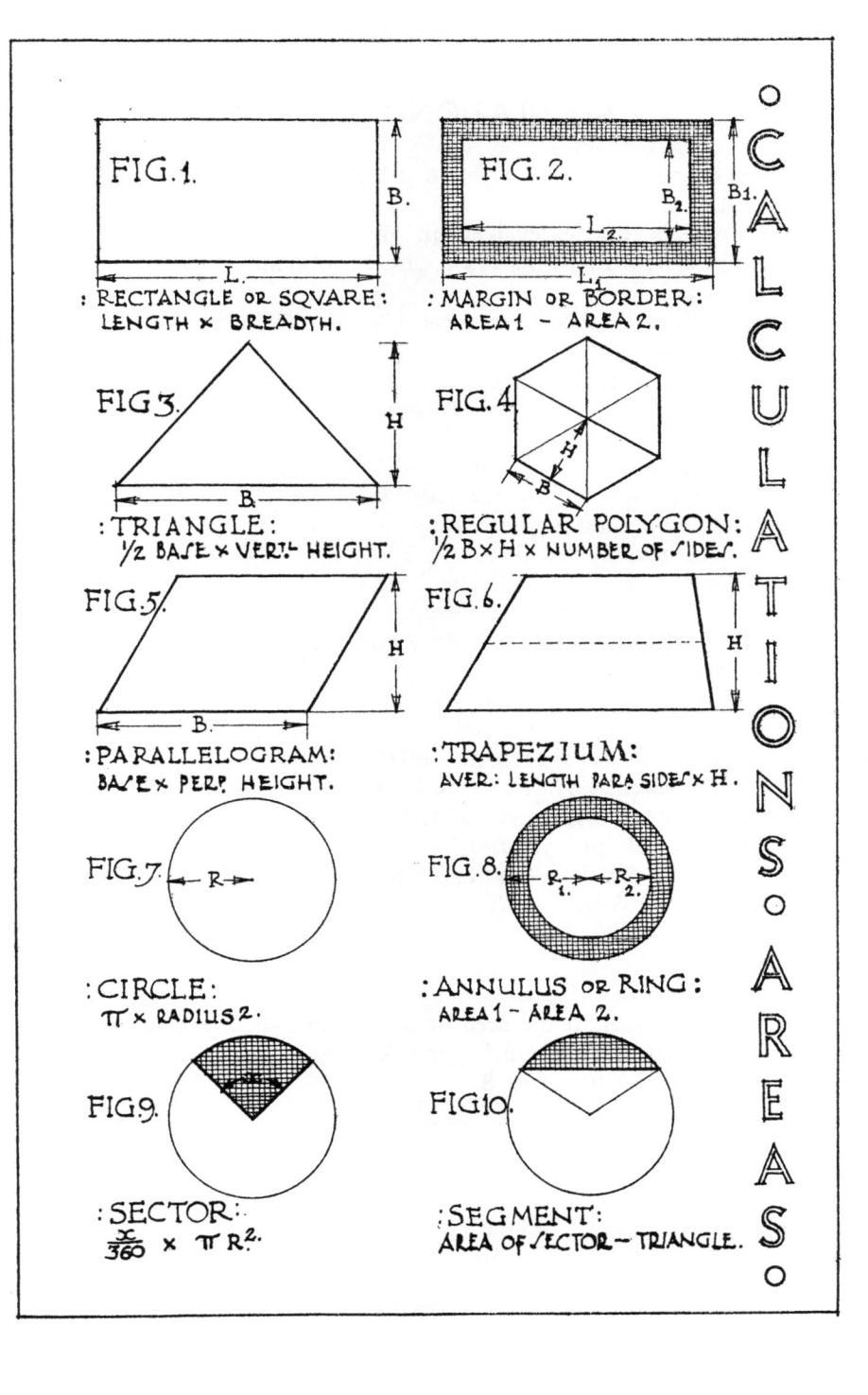

CALCULATIONS OF AREAS

FIG. 1.
: RECTANGLE OR SQUARE :
LENGTH × BREADTH.

FIG. 2.
: MARGIN OR BORDER :
AREA 1 − AREA 2.

FIG. 3.
: TRIANGLE :
½ BASE × VERT.^{L} HEIGHT.

FIG. 4.
: REGULAR POLYGON :
½ B × H × NUMBER OF SIDES.

FIG. 5.
: PARALLELOGRAM :
BASE × PERP. HEIGHT.

FIG. 6.
: TRAPEZIUM :
AVER: LENGTH PARA SIDES × H.

FIG. 7.
: CIRCLE :
π × RADIUS^2.

FIG. 8.
: ANNULUS OR RING :
AREA 1 − AREA 2.

FIG. 9.
: SECTOR :
$\frac{x}{360}$ × π R^2.

FIG. 10.
: SEGMENT :
AREA OF SECTOR − TRIANGLE.

CALCULATION OF VOLUMES

THE volume of a solid is obtained from three dimensions: length, breadth, and height.

Cubing is used extensively in building work for the measurement of quantities of materials, also for obtaining the cubical contents of buildings.

As with calculations for areas these few formulae are given for reference only.

Volume of cube, Fig. 1:
  LENGTH × BREADTH × HEIGHT.

Volume of prism, Fig. 2:
  AREA OF BASE × HEIGHT.

This applies to any parallel sided solid, i.e. one having a constant section throughout its length.

Volume of cone, Fig. 3:
  AREA OF BASE × ⅓ HEIGHT. $\pi R^2 H/3$.

Area of curved surface of cone:
  π × RADIUS OF BASE × SLANT HEIGHT. πRL.

Volume of pyramid, Fig. 4:
  AREA OF BASE × ⅓ HEIGHT.

Volume of cylinder, Fig. 5:
  AREA OF BASE × HEIGHT. $\pi R^2 H$.

Area of curved surface of right cylinder:
  PERIMETER × HEIGHT. $2\pi RH$.

Volume of oblique cylinder:
  AREA OF BASE × VERTICAL HEIGHT.

Volume of sphere, Fig. 7:
  $\frac{4}{3}\pi R^3$.

Area of surface of sphere:
  $4\pi R^2$ (equal to the area of the curved surface of a cylinder that would contain the sphere), Fig. 7.

Volume of frustum, Fig. 8:
  $\frac{1}{3} H (A_1 + A_2 + \sqrt{A_1 \times A_2})$.

CALCULATIONS · VOLUMES

FIG 1
CUBE
LENGTH × BREADTH × HEIGHT.

FIG. 2
:PRISM:
AREA OF BASE × HEIGHT.

FIG 3
CONE
AREA OF BASE × $\frac{1}{3}$ HEIGHT

FIG 4
:PYRAMID:
AREA OF BASE × $\frac{1}{3}$ HEIGHT

FIG 5
:RIGHT CYLINDER:
AREA OF BASE × HEIGHT.

FIG 6
OBLIQUE CYLINDER:
AREA OF BASE × VERT. HEIGHT.

FIG. 7
:SPHERE:
$\frac{4}{3} \pi R^3$.

FIG 8
:FRUSTUM.
$\frac{1}{3} H (A_1 + A_2 + \sqrt{A_1 \times A_2})$

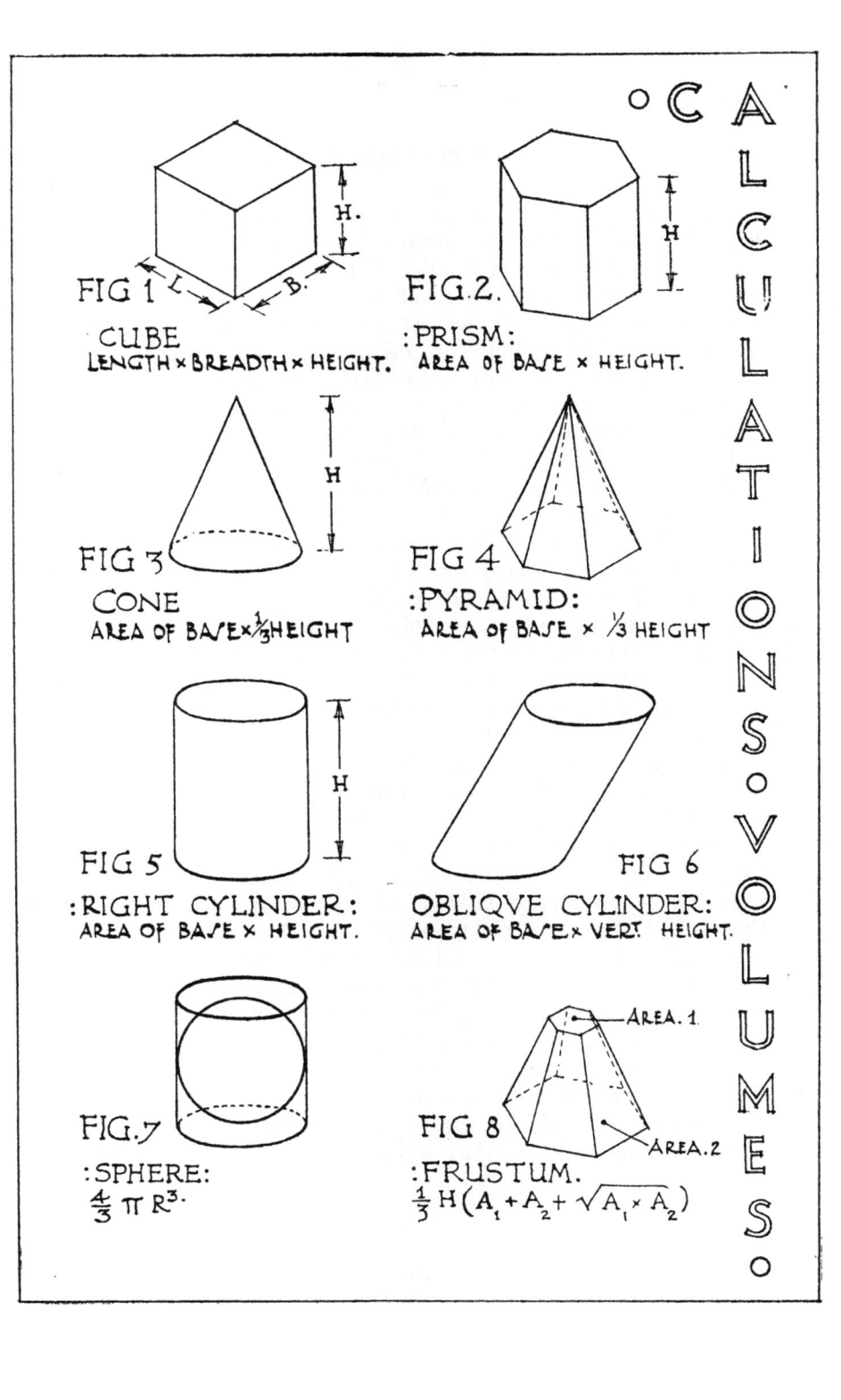

SIMILITUDE

SIMILAR figures have their angles equal and their sides proportional. It is often necessary to alter the size of an irregular figure, such as a moulded section, proportionally. This can be done graphically by a method known as radical projection or the principle of similitude.

It is required to reduce the size of the cavetto mould enclosed in the rectangle ABCD, Fig. 1, to one-half (i.e. the dimensions, not the area; dividing the sides by two divides the area by four).

Produce the base CD to any conveniently situated pole point P. The required section is to be half the original size, therefore make PD^1 half PD. Radiate ABC and D to P.

The vertical A^{1}D^1 is half the height of AD and PA^1 is half PA, so that the rectangle A^{1}B^{1}C^{1}D^1 is a similar figure to ABCD. In this simple case a new striking centre for the curve could be found in a similar manner, but in Fig. 1 a series of points on it have been taken to make the method clearer. The points taken on the original curve are radiated to the pole point and their location on the new section obtained by another projection of the point on to the outside of the enclosing rectangle. X–X is the projection of the point in the original section and X^1–X^1 gives its position on the required section.

It is immaterial where the pole point is situated, as shown by Fig. 2, both solving the same problem. Again, one of the radial lines is divided according to the proportional size of the required section.

The method adopted in Fig. 3 is termed inverse similitude, and the pole point is situated between the two sections.

In Fig. 4 a section is reduced in one dimension only, namely, the amount of projection, by one-quarter. Draw AE at any angle, making its length equal to 3/4 BA. Join BE and make the other projection lines parallel to it. With centre A rebat these points on to the horizontal AB^1, dropping them down until they meet the corresponding lines carried across horizontally.

In Fig. 5 a pole point has been chosen actually at point D so that the original proportioning is done on line DC, giving in this case DC^1 as half DC. The drawing is complicated, and the sole merit of the method is the economy of space.

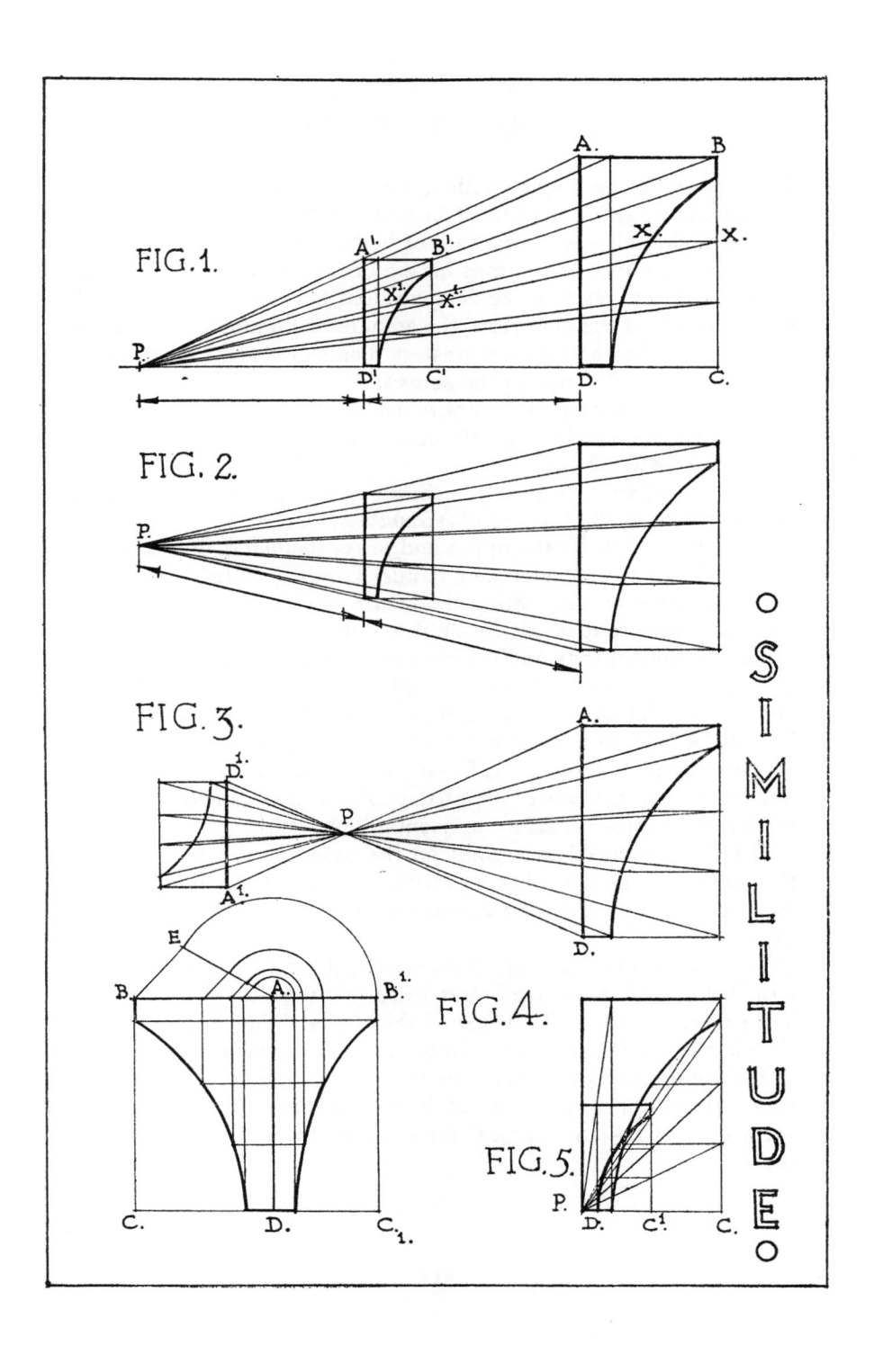

FIG. 1.

FIG. 2.

FIG. 3.

FIG. 4.

FIG. 5.

O SIMILITUDE O

ENTASIS OF COLUMN

The entasis is the slight swelling that is given to the shaft of a column to counteract the optical illusion of hollowness. The amount is only slight and any appearance of bellying should be avoided.

Classic examples of columns have a definite proportion between their height and their lower diameter and between the upper and lower diameters which governs the amount of diminish. The lower diameter of the shaft is the greatest diameter and at no place is this width exceeded due to the entasis.

The entasis may be on the entire length of the shaft or the lower portion may be parallel, with the entasis and diminish on the upper two-thirds only.

The first of the two methods shown is the one most often used and is known as the *Concoid of Nicomedes*.

Let AB and CD be the upper and lower diameters of a column, Fig. 1. With C as centre and radius X, equal to half the lower diameter, draw an arc cutting the centre line of the column in E. Join CE and produce, either until it cuts the base line AB produced, or until it cuts any conveniently placed line parallel to the centre line as at GH. Divide EF into any number of equal parts and divide GH in a similar manner, joining the points by radial lines 1–1¹, 2–2¹, etc. With centres 1, 2, 3, etc., and radius X, draw arcs cutting the lines in J, K, L, etc., which are points in the curve. The entasis on the other side of the column is obtained by transferring the widths across on horizontal lines.

If CE is produced to a point on the base line the divisions on the centre line of the column radiate direct from this point. This method may save a certain amount of work but space rarely permits it.

The second method, Fig. 2, gives a slightly fuller curve. On the base line draw a semicircle representing half the plan of the column and project on to it the width of the upper diameter as D to E. Divide the arc EO into any number of equal parts and the height of the column into a similar number of parts. The points in the curve are at the intersection of horizontal lines drawn from the vertical divisions and vertical lines drawn from the divisions on the arc.

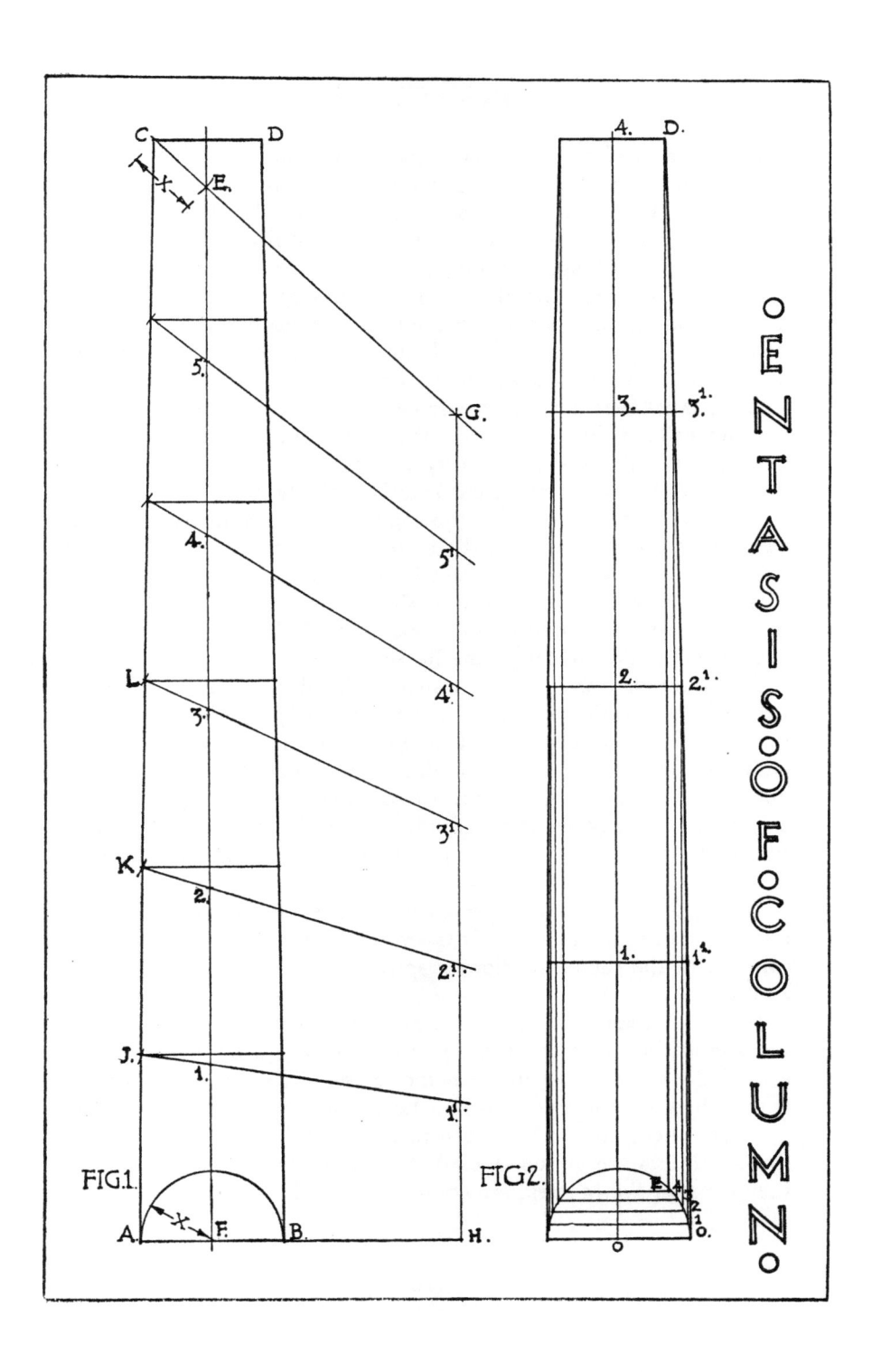

FIG.1.

FIG.2.

ENTASIS OF COLUMN.

SPIRAL CURVES

A SPIRAL, as a plane figure, is a locus of a point revolving about a centre whilst at the same time receding from it in a constant manner. The curve may also rise continuously throughout its length, with a constant ratio between rise and distance travelled.

Many forms of spiral curves are used in architecture, most of them based on geometrical principles, although some are executed as freehand ornament in much the same way as carving.

Trusses, console brackets, column caps, handrail terminations, and curtail steps are examples of spirals.

A LOGARITHMIC SPIRAL is shown in Fig. 1; the actual line is drawn freehand through a series of predetermined points. From O draw twelve equally-spaced radial divisions and on the first of these mark off AO of any convenient length as the first radius vector.

Bisect AO in C and using this centre draw a semicircle cutting the next division line in D, the second point in the curve. Bisect DO, giving C^1 as the next centre; a semicircle from this centre with DC^1 as radius cuts the next division line in E, a third point in the curve. Repeat this operation to find the series of centres C^2, C^3, etc., and further points in the curve.

The ARCHIMEDEAN SPIRAL is also a freehand line drawn after first determining a series of points in the curve, Fig. 2. From centre C draw a series of equally-spaced angular divisions; twelve are usually sufficient and are easy to obtain, each being 30°. Divide the vertical OC into the same number of equal divisions and using C as centre draw a series of arcs, point 1 to radial division 1, point 2 to radial division 2, etc., giving a series of points in the required curve.

In Fig. 2 the curve makes one complete revolution and is termed a spiral of one convolution. If the vertical divisions were twice the number of the radial divisions a spiral of two convolutions would result.

In Fig. 3 a spiral is shown based on quadrants of circles. On a horizontal line mark off ten equal divisions and draw the centre eye of the spiral equal in diameter to two divisions. Construct a square beneath the first division, the angular points of which, 1, 2, 3, 4, are successive striking centres. The points of separation of the quadrants are at the sides of the square produced, which, as they connect the centres, are common normals.

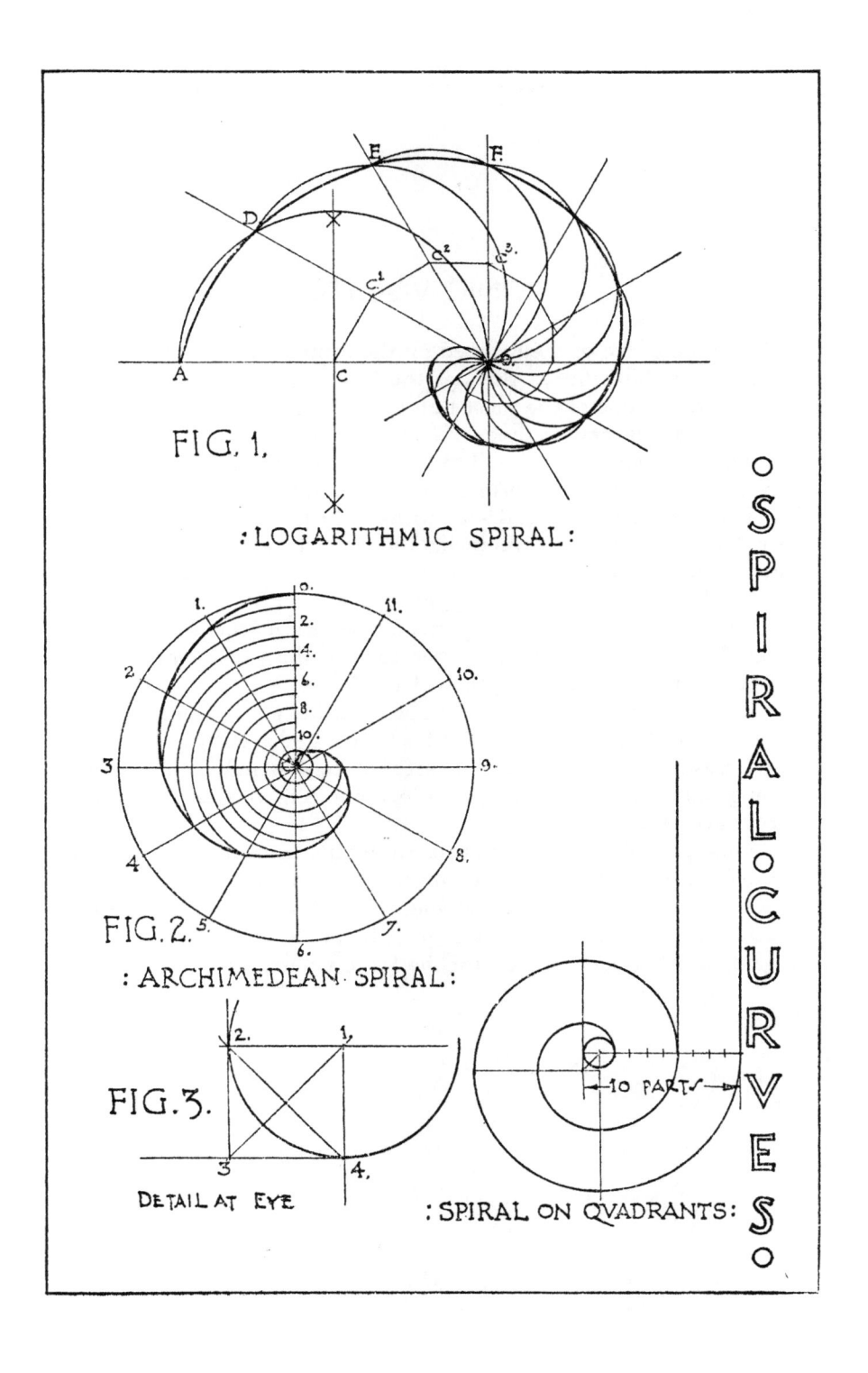

FIG. 1.

:LOGARITHMIC SPIRAL:

FIG. 2.

:ARCHIMEDEAN SPIRAL:

FIG. 3.

DETAIL AT EYE

←10 PARTS→

:SPIRAL ON QUADRANTS:

SPIRAL␣CURVES

IONIC VOLUTE

THE spiral curve or volute used on the capitals in the Greek Ionic order of architecture is perhaps the best-known use of the spiral and certainly one of the most beautiful of curves.

There are several methods of constructing the curve, each giving a slightly different result. The method illustrated here is known as GIBB'S RULE and is a form of construction in very general use.

The complete curve is drawn within a rectangle 8 units in height by 7 units in width. Commence the construction by drawing the eye having a diameter of 1 unit and its centre horizontally between 3 and 4 and vertically above point 3.

Within the eye draw a square, as shown in the enlargement, Fig. 2, bisect the sides and divide each of the two diameters into six equal parts, numbering them 1 to 12.

Referring back to Fig. 1, it will be seen that the first arc is drawn with centre 1 and radius 1–8[1] and is terminated at A by a horizontal line drawn from centre 2. The second arc is struck from centre 2 using radius 2–A and terminates in point B, a vertical line drawn from centre 3.

This procedure is continued throughout, using each of the twelve centres in turn. The volute is one of three convolutions, each convolution consisting of four arcs. The centres for the two inside turns of the volute are 5, 6, 7, 8, and 9, 10, 11, 12.

For another construction for the Ionic volute, see page 120.

FIG. 1.

FIG. 2.

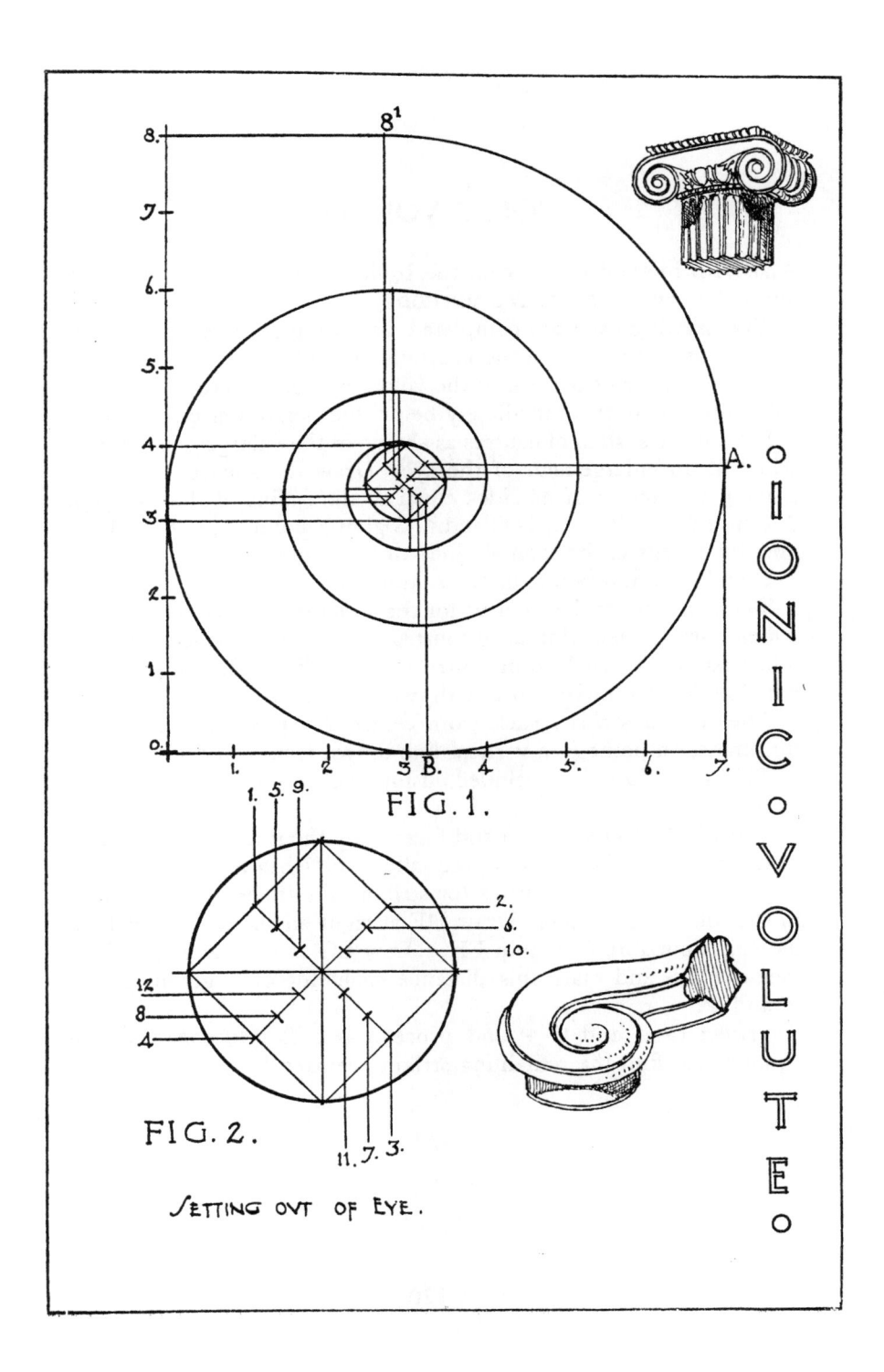

SETTING OUT OF EYE.

IONIC VOLUTE

IONIC VOLUTE

A SECOND method of drawing the Ionic volute is shown in Fig. 1 and is known as GOLDMAN'S METHOD.

The drawing has been completed, including the line of the fillet, which was not shown on the previous example.

The diameter of the eye of the volute is $\frac{1}{8}$ of its total height and its position is fixed immediately below the horizontal centre line.

Set out the system of centres as shown in the enlargement, Fig. 2, which is an enlargement of the eye. Draw the square 1, 2, 3, 4, making the length of its sides equal to the radius of the eye, join 2–C and 3–C. Trisect 1–C and 4–C giving points 5, 9, 8, and 12, and by means of horizontal lines drawn from these points on to 2–C and 3–C find points 6, 10, 7, and 11.

These are the twelve centres for the arcs forming the outer curve which now becomes similar in construction to the previous example. The first arc is struck from centre 1 with radius 1–D and is terminated at A by a horizontal line drawn from centre 2.

The second arc is struck from centre 2 using radius 2–A and terminates in point B, a vertical line drawn from centre 3.

This procedure is continued, using each of the twelve centres in turn.

The width at DE of the band formed by the inner curve is equal to the radius of the eye and gradually diminishes towards the eye.

The new system of centres for striking the inner curve is found in the following manner: Draw DF at right angles to DC, making its length equal to C–1, join FH. Draw EG at right angles to DC, measure EG and mark this distance each side of C giving points 1^x and 4^x.

Trisect $C-1^x$ and $C-4^x$ and proceed as before to construct the squares and find the remaining striking centres.

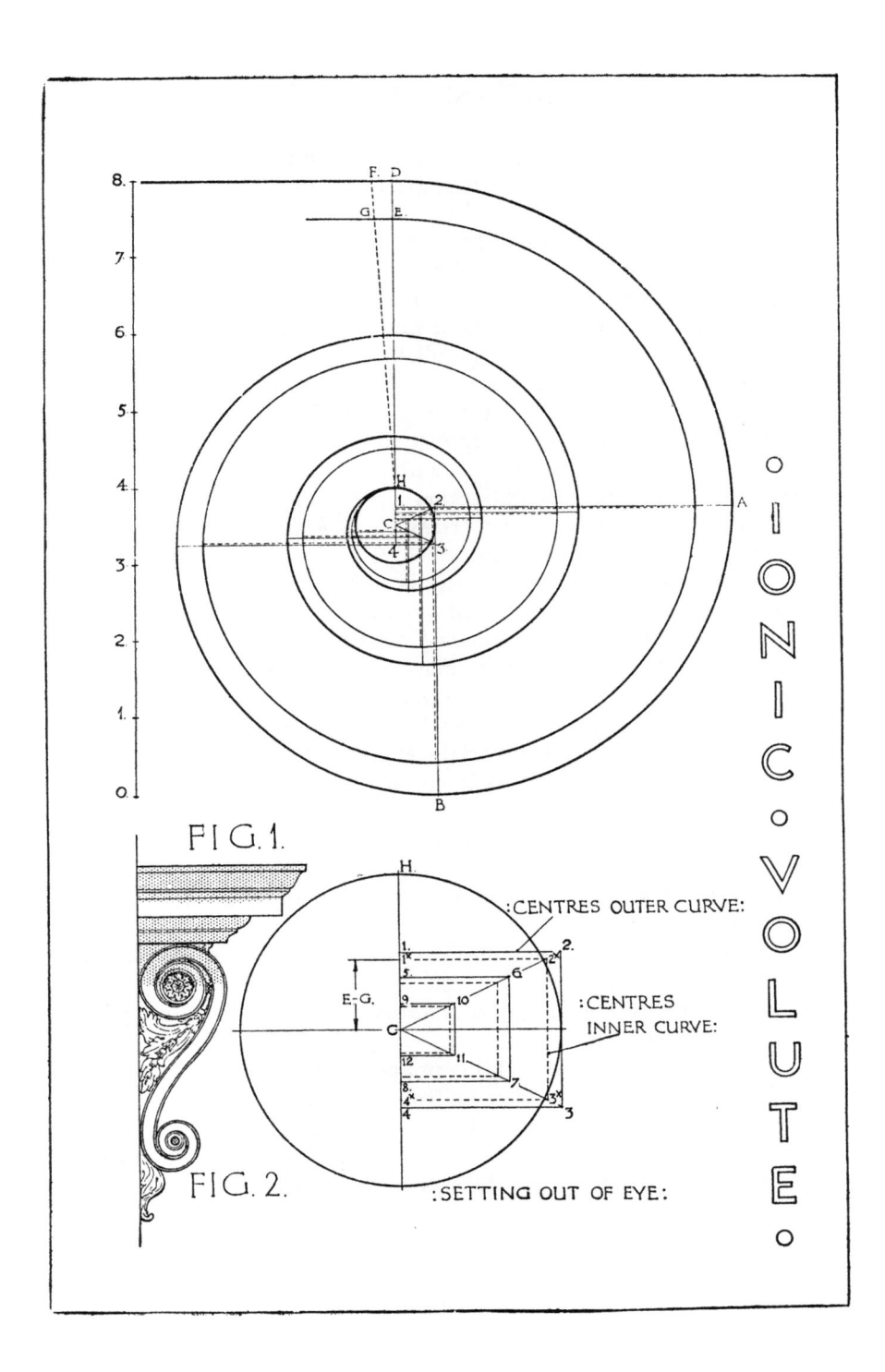

FIG.1.

FIG.2.

CENTRES OUTER CURVE:

CENTRES INNER CURVE:

SETTING OUT OF EYE:

IONIC·VOLUTE

GEOMETRICAL SOLIDS

A SOLID has three dimensions: length, breadth, and depth. A few only, used in the projection examples that follow, are represented here.

A CUBE is a solid bounded by and contained by six equal squares.

A PARALLELOPIPED is a solid contained by six quadrilateral figures each opposite pair of which are parallel.

A PRISM is a solid the two ends of which are parallel plane figures and whose sides are parallelograms. A prism whose ends are at right angles to its sides is a *right prism* and one whose ends are not at right angles to its sides is an *oblique prism*. Prisms are usually named according to the shape of their ends. Three right prisms are shown in elevation and plan, Figs. 1 to 3. A *square prism*, Fig. 1, has its end section a square. A *triangular prism*, Fig. 2, has its end section a triangle. A *hexagonal prism*, Fig. 3, has its end section a hexagon.

A PYRAMID is a solid whose base is a plane figure and whose sides are triangular. The elevation of a pyramid is always a triangle. As with prisms, pyramids are named according to the form of their base, and those illustrated in Figs. 4, 5, and 6 are respectively a *square*, a *triangular*, and a *hexagonal pyramid*.

The axis of a pyramid is a line drawn from the apex to the centre of the base. If the axis is at right angles to the base it is termed a *right pyramid* and if the axis is not at right angles to the base it is termed an *oblique pyramid*.

A CONE, Fig. 7, is a solid generated by the revolution of a right-angled triangle about one of its sides, containing the right angle, as a fixed axis. The base and plan of a cone is a circle and its elevation a triangle.

A *truncated* cone or pyramid is one whose apex has been cut off by a plane section.

A FRUSTUM is a portion of a solid contained between two planes parallel to the base.

A SPHERE, Fig. 8, is a solid contained within a single surface, all points in which are equidistant from the centre. The elevation of a sphere viewed from any direction is a circle.

A CYLINDER, Fig. 9, is a solid formed by the revolution of a rectangle about one of its sides as a fixed axis.

FIG.1. FIG.2. FIG.3.

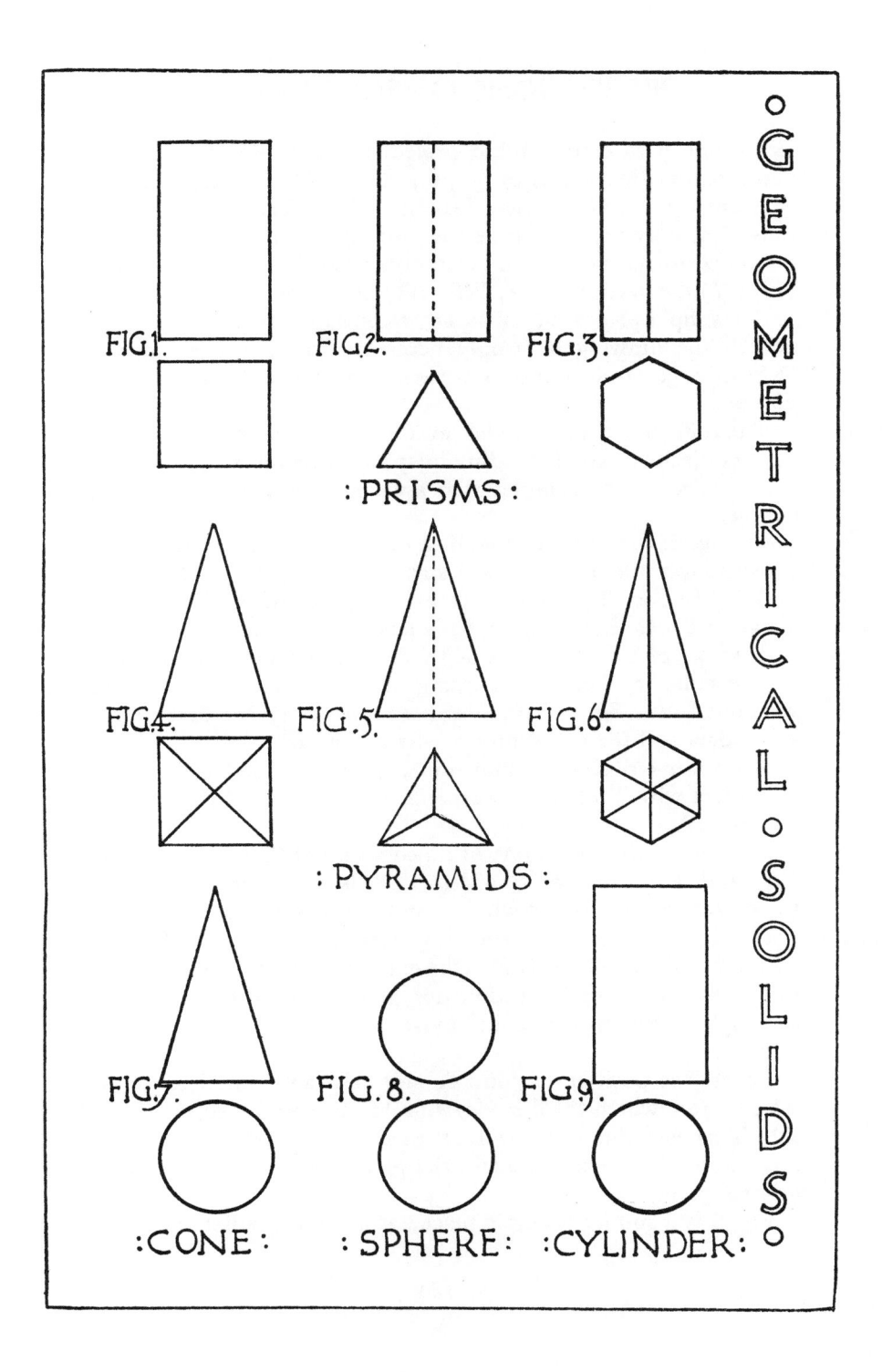

: PRISMS :

FIG.4. FIG.5. FIG.6.

: PYRAMIDS :

FIG.7. FIG.8. FIG.9.

: CONE : : SPHERE : : CYLINDER :

GEOMETRICAL SOLIDS

PROJECTIONS OF SOLIDS: CUBE

In dealing with orthographic projection (page 40) reference was made to imaginary planes of projection. These planes must be constantly referred to when dealing with projections of solids because it is only possible to fix the position of an object in space by reference to its location in relation to: (*a*) The horizontal plane, H.P.; (*b*) the vertical plane, V.P.; (*c*) the side vertical plane, S.V.P.

For simple projections it is not essential to refer to the S.V.P. and in the drawing it is only necessary to show the hinge line (XY in Figs. 1–3 opposite) between the vertical and horizontal planes.

The difficulty is likely to be not the drawing but the visualizing of the object as a solid standing in space. It is not therefore superfluous to attempt examples that appear, as drawings, to be very simple.

In Fig. 1 the cube is standing on H.P. and in front of V.P., consequently the elevation will stand on XY and the plan is clear of XY. In Fig. 3 the cube has one edge touching V.P. and it is above H.P. In Fig. 2 the cube is suspended in space above H.P. and in front of V.P., therefore neither elevation nor plan touches XY.

The cube in Fig. 4 is standing on H.P. and in front of V.P. with one face at 30° to V.P. The plan must be drawn first from the given data and the elevation projected from it.

The reverse position is shown in Fig. 6; the cube has one surface in contact with V.P. and is suspended above H.P. with one face at 30° to H.P.

A cube standing with one of its edges on H.P. at right angles to V.P. and with one face at 45° to H.P. is shown in Fig. 5. Here the elevation must be drawn first and the plan projected from it.

In Fig. 7 the cube is standing with one corner on H.P. and with one side at 30° to H.P. One edge is in contact with V.P. In this case neither the elevation nor plan can be drawn direct and a simple view of the cube must first be constructed (shown dotted).

Locate the position of point A, side AD, and complete the plan ABCD from which the first elevation can be completed, A^1, B^1, C^1, D^1. This elevation can now be turned about point A^2 into its new position at the given angle and the plan projected from it and the original plan.

Fig. 8 is a similar example placed at a slightly different angle.

SIMPLE PROJECTIONS. CUBE.

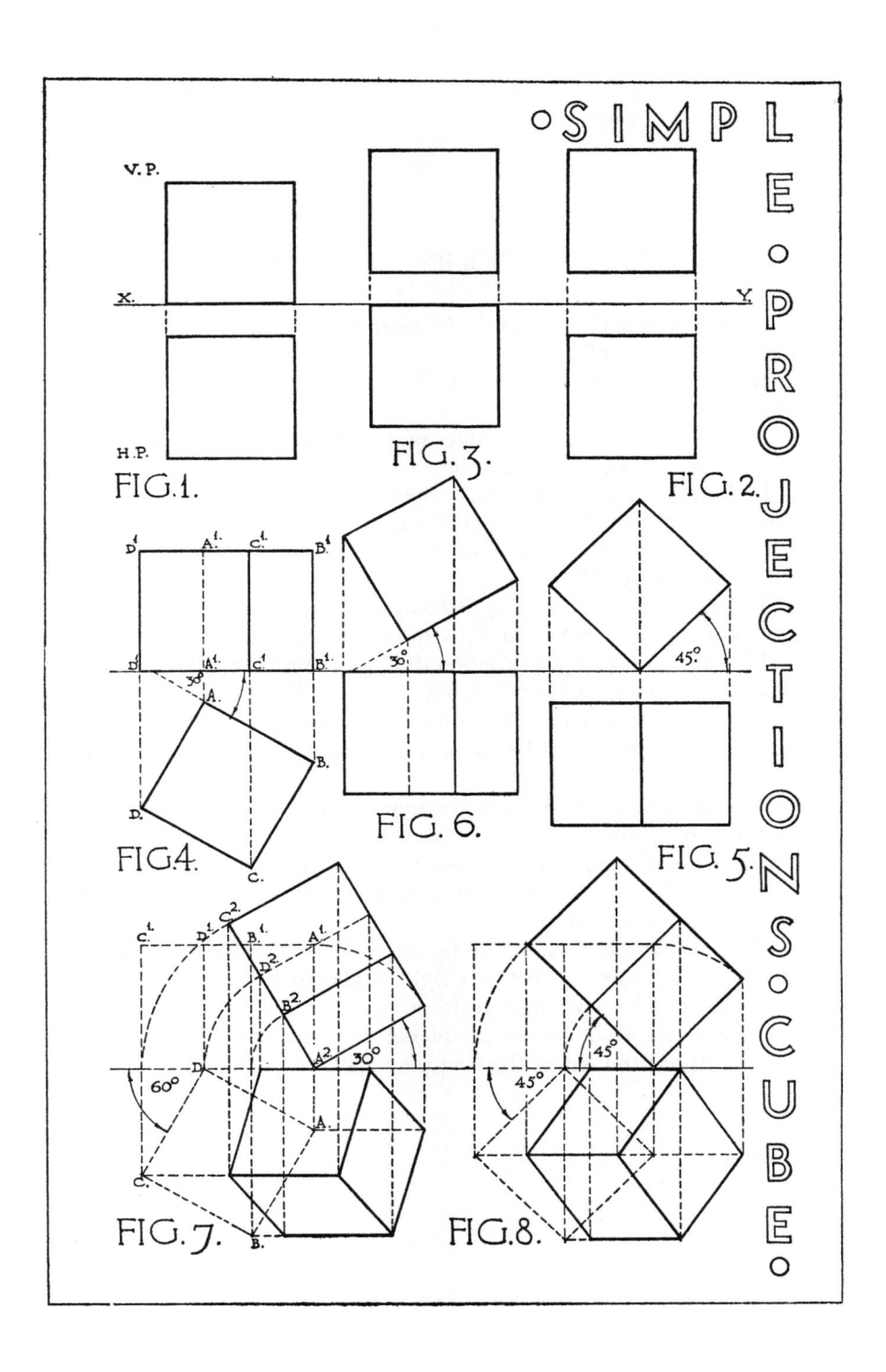

PROJECTIONS OF SOLIDS: TRIANGULAR PRISM

THE projections of the triangular prism are treated here in a similar manner to those of the cube.

In Fig. 1 it is standing on H.P. with one face parallel to V.P. and in front of it.

In Fig. 2 one end is in contact with V.P. and one side is parallel to H.P. but suspended above it.

In Fig. 3 the prism has one angular edge in contact with V.P.; the base is above H.P. but parallel to it.

In Fig. 4 the prism is standing on H.P.; it is in front of V.P. with one side at 30° to V.P.

The inclined view in Fig. 5 has one base edge in contact with H.P. and at right angles to V.P. and one side at 30° to the H.P.

The elevation and plan are first drawn in their simplest form, and then the elevation is turned about the line AB into the new angle.

In Fig. 6 the prism is standing with one side above H.P. and parallel to it, one corner being in contact with V.P. and one angular edge at 30° to V.P.

Here the angle of 30° is in relation to the V.P. so it is the plan that is turned about the fixed point A into its new position.

Fig. 7 is a triangular prism standing with one corner on H.P. and its base at 30° to the H.P. It is standing in front of V.P. with one side at right angles to it.

ABC is a simple plan of the figure, and $A^1B^1C^1$ the plan turned about point C to make B^1C^1 at right angles to V.P. The elevation is turned into its new angle from point A^2.

In this example, of course, the plan $A^1B^1C^1$ could have been drawn direct without the aid of the first plan ABC.

FIG. 1. FIG. 2. FIG. 3. FIG. 4. FIG. 5. FIG. 6. FIG. 7.

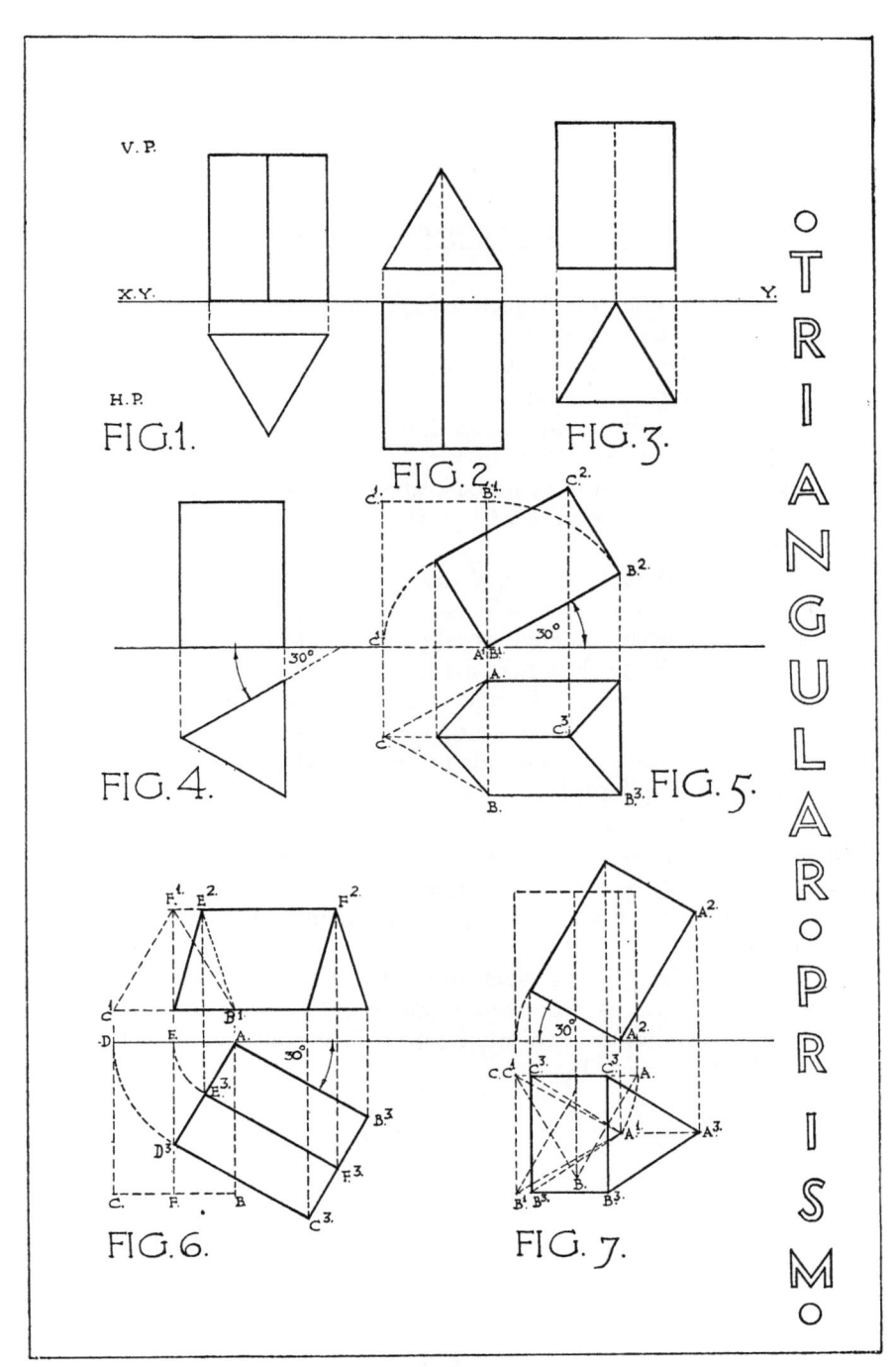

TRIANGULAR PRISM

PROJECTIONS OF SOLIDS: HEXAGONAL PRISM

FURTHER examples of projections are provided by the hexagonal prism in various positions.

In Fig. 1 the prism is standing on H.P.; it is in front of V.P. with one face parallel to it.

Fig. 3 shows the prism above the H.P. with one side parallel to it and one end in contact with V.P.

In Fig. 2 the prism has its base above and parallel to H.P. One angular edge is in contact with V.P. and two adjacent sides are at 30° to V.P.

The inclined view in Fig. 4 has one base edge on H.P. and at right angles to V.P.; one side is at 30° to H.P. It is standing in front of V.P. with one angular edge parallel to it.

The plan ABCDEF is first drawn and the elevation standing on H.P. projected from it. As one side is to be inclined at 30° to the H.P. the elevation can now be turned about the fixed line BC into this required position. The plan is projected from the new elevation and the original plan.

The example, Fig. 5, is of a similar prism standing with one corner on H.P. and its base inclined at 30° to the H.P. It is standing in front of V.P. with one face parallel to it.

The method of construction is the same as in the previous example.

In Fig. 6 the prism is standing with one side on H.P., it is in front of V.P. and has one angular edge at 30° to V.P.

In Fig. 7 it stands with one edge on H.P., in front of V.P., and one side at 45° to V.P.

FIG. 1. FIG. 2. FIG. 3. FIG. 4. FIG. 5. FIG. 6. FIG. 7.

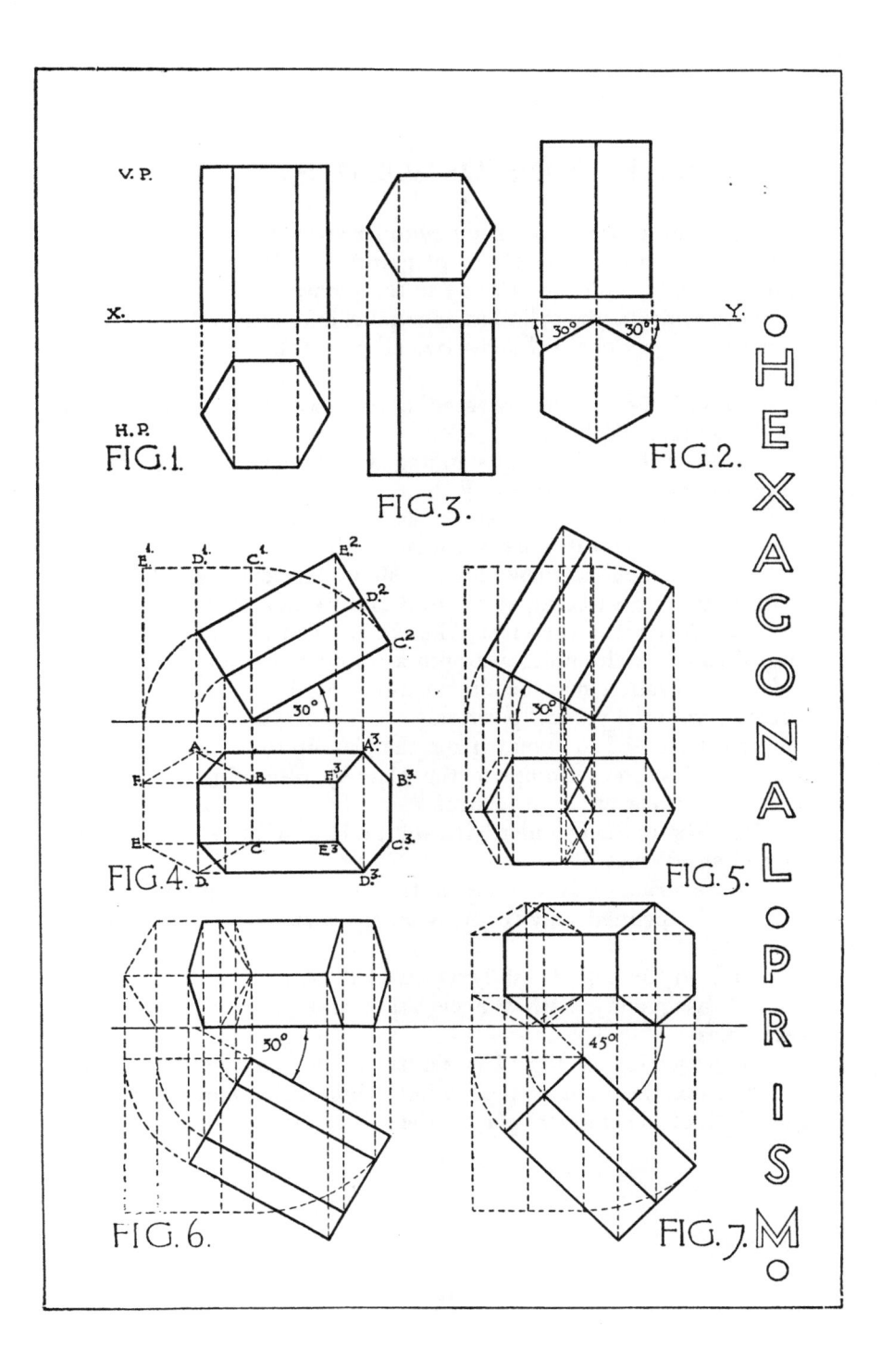

HEXAGONAL PRISM

PROJECTIONS OF SOLIDS: CYLINDER

THE two direct views of a right cylinder are a rectangle and a circle either in elevation and plan or plan and elevation according to the position of the solid in relation to the planes of projection.

As the sides of the cylinder are parallel and both ends circles, the inclined projections will show two ellipses either in plan or elevation.

In Fig. 1 the cylinder is standing in space with its axis vertical to the horizontal plane.

In Fig. 2 the cylinder is standing with its base on H.P. and its side in contact also with the V.P.

The reverse position is shown in Fig. 3 where the cylinder is lying on H.P. with its base in contact with V.P.

The first inclined view, Fig. 4, shows the cylinder in contact with H.P. its base being at 45° to it and its axis parallel to V.P.

Draw the circle representing the normal plan view of the cylinder and divide the circumference into a number of equal parts. The inclined elevation may be drawn direct at the required angle or rebated from the vertical position.

Project the division points from the plan up to XY and on to the inclined elevation. Complete the plan by projecting down from the inclined elevation and across from the original plan.

Fig. 5 shows the cylinder lying on the H.P. with its base in front of and at 60° to V.P.

The construction is identical to the previous example but as the cylinder is inclined only to the V.P. the rectangular view will be the plan.

In Fig. 6 the cylinder is lying on H.P. with its axis parallel to V.P., consequently both the elevation and plan will be similar rectangles.

The projection in Fig. ¶7 is similar to that in Fig. 4 with the exception of the angle of inclination, which being steeper reduces the width of the elliptical plan of the end of the cylinder.

V.P.
X.   Y.

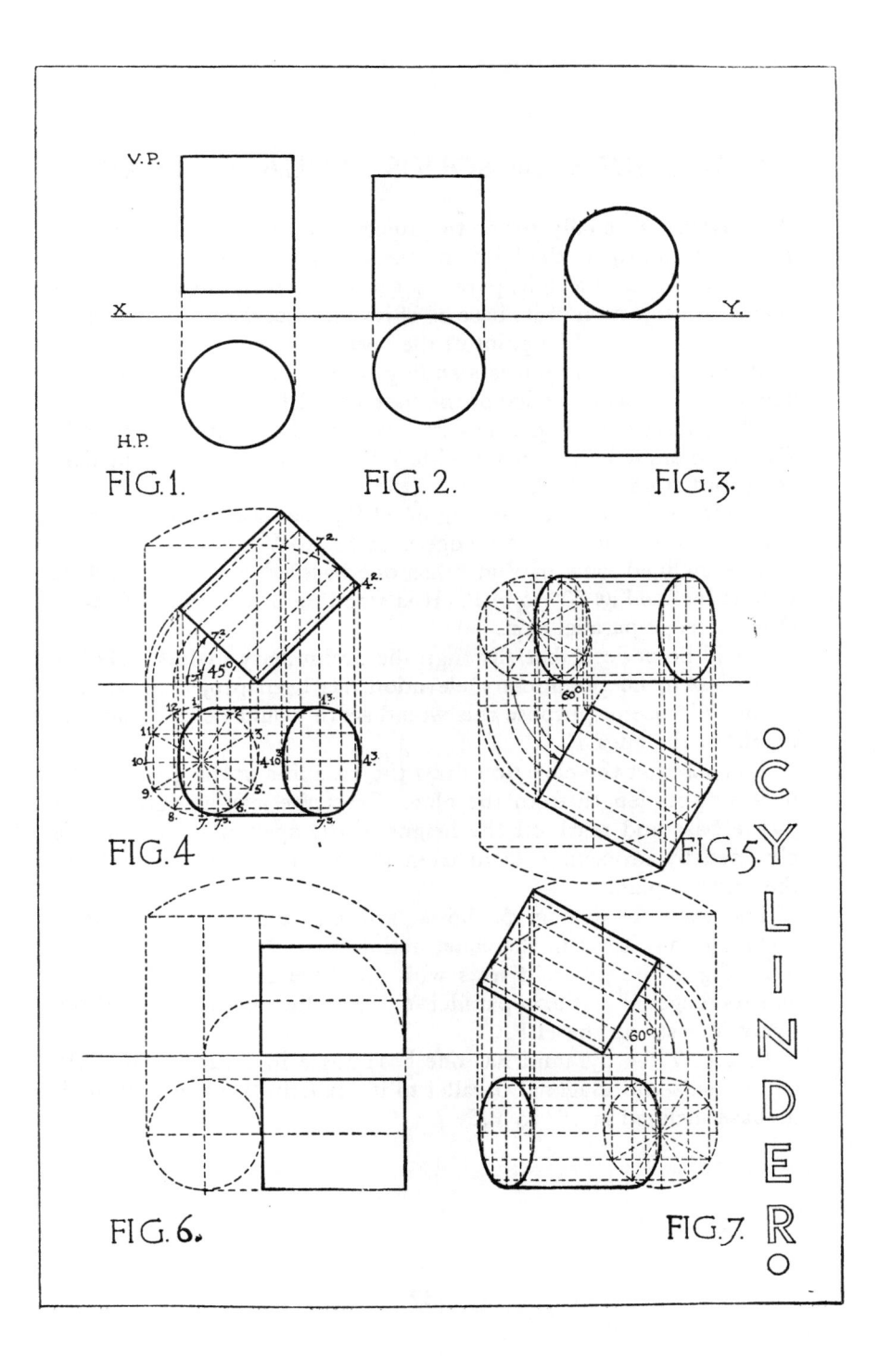

H.P.

FIG. 1.   FIG. 2.   FIG. 3.

FIG. 4   FIG. 5.

FIG. 6.   FIG. 7.

CYLINDER O

PROJECTIONS OF SOLIDS: SQUARE PYRAMID

A PYRAMID is usually easier to project than a prism as it is only necessary to project the base and the point of the apex.

The position of a right pyramid can be fixed by the angle to the planes of projection of its base or of its axis, i.e. the line connecting the apex to the middle point of the base.

In Fig. 1 the pyramid is standing on the H.P., it is in front of the V.P. and has one edge of the base parallel to it.

The pyramid in Fig. 2 has its base parallel to and above H.P. One base angle is in contact with V.P. and the two adjacent base sedge are at 45° to V.P.

In Fig. 3 the solid is standing on H.P., one base angle is in contact with V.P. and one base edge is at 60° to V.P.

The inclined view in Fig. 4 has one base edge on H.P. and its base is inclined at 60° to H.P. It is standing in front of V.P. with the base edge parallel to it.

It will be noticed that although the preliminary plan ABCD has been drawn, no preliminary elevation has been projected from it as the only useful purpose this would serve would be to obtain the height of the apex E.

To construct the elevation draw the base line at the given angle, obtaining its length from the plan. Erect the axis at right angles to the base and mark off the height of the apex E^1. The plan is obtained by projecting down from the elevation and across from the original plan.

The example in Fig. 5 shows how enclosing the figure in a rectangle may assist in the construction.

In Fig. 6 the pyramid rests with one base angle touching V.P. and its diagonal and axis parallel to V.P. One angular edge of the pyramid rests on the H.P.

In Fig. 7 the pyramid has one base angle in contact with H.P. and its axis and diagonal parallel to it. It is in front of V.P. with its base inclined at 45° to V.P.

SQUARE PYRAMID

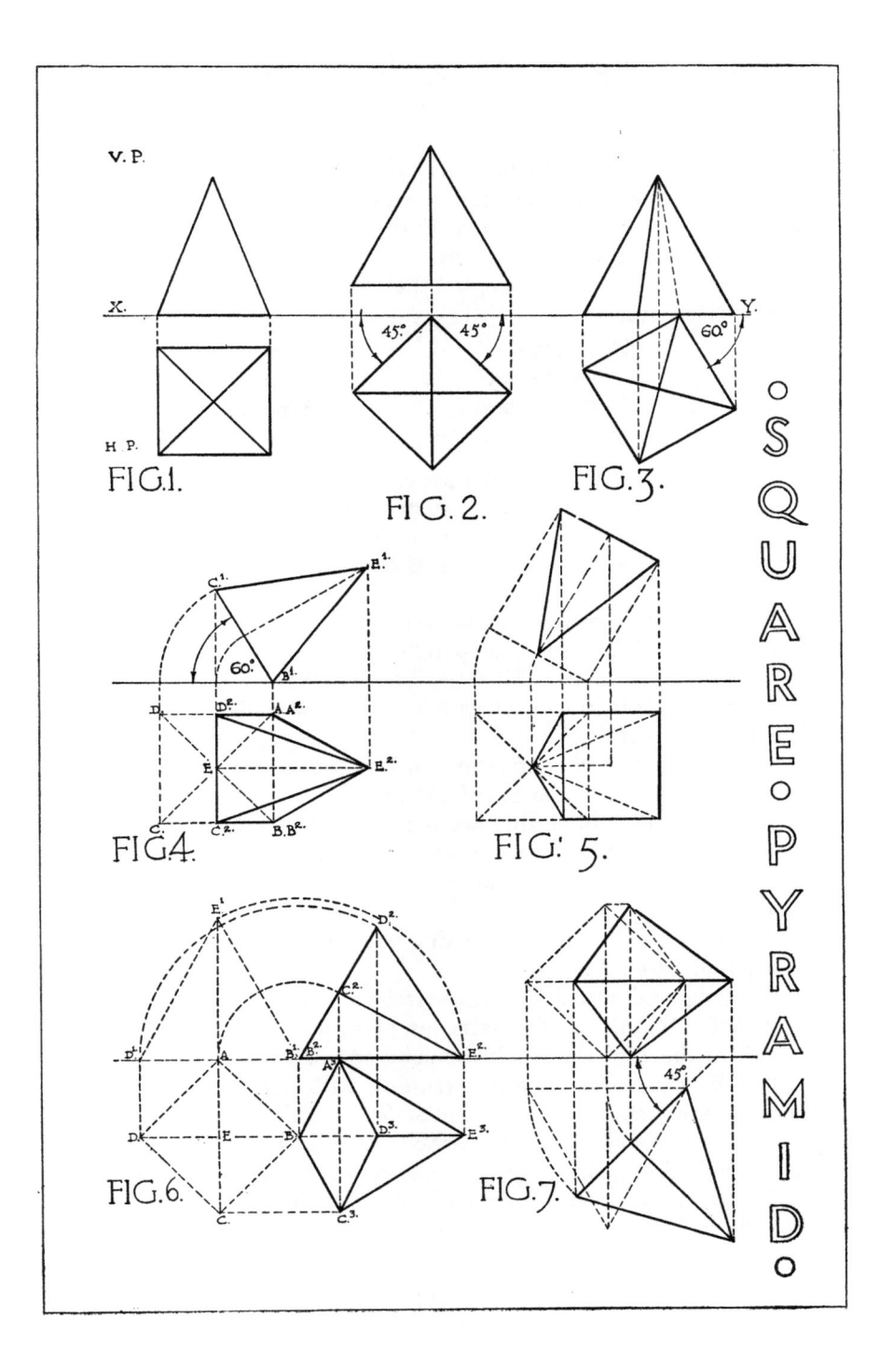

FIG.1.
FIG.2.
FIG.3.
FIG.4.
FIG.5.
FIG.6.
FIG.7.

PROJECTIONS OF SOLIDS: TRIANGULAR PYRAMID

THE triangular pyramid has each of its four faces triangles. If it is a *right pyramid* three of these are equal, but the fourth, the base, is not necessarily equal to them. If the figure has four faces of equal area it is termed a TETRAHEDRON.

Other regular solids having all their faces equal in area and similar in shape are:

HEXAHEDRON	. .	6 faces.
OCTAHEDRON	. .	8 faces.
DODECAHEDRON	. .	12 faces.
ICOSAHEDRON	. .	20 faces.

If a regular solid is within a sphere each of the angular points of the solid is in contact with the surface of the sphere.

The projection examples given here are similar to those on previous pages and the methods of construction should need little explanation.

In Fig. 1 the pyramid is standing on H.P., it is in front of V.P. and has one edge of the base parallel to it.

The pyramid in Fig. 2 has its base parallel to and above H.P. One base angle is in contact with V.P. and two adjacent base edges are at 60° to V.P.

In Fig. 3 the solid is standing on H.P., it is in front of V.P. and one base edge is at an angle of 15° to V.P.

It should be noted that as the sides of the figure are equal the true plan of the base is always an equilateral triangle, consequently the plan is readily drawn after finding the position of one side.

The example in Fig. 4 is resting with one edge on H.P. and its axis parallel to V.P.

The edge AC remains in a vertical plane so that the new elevation $A^2C^2D^2$ will be the same as elevation $A^1C^1D^1$. The true length of the edge on which the solid stands is the line A^3D^3.

Fig. 5 shows the pyramid standing with one corner on H.P., with its axis parallel to V.P., and at 60° to H.P.

In Fig. 6 the solid is turned about its apex until the axis makes an angle of 30° with the H.P. whilst remaining parallel to V.P.

The pyramid in Fig. 7 is standing with its apex on H.P., its axis parallel to V.P. and at 30° to H.P.

TRIANGULAR PYRAMID

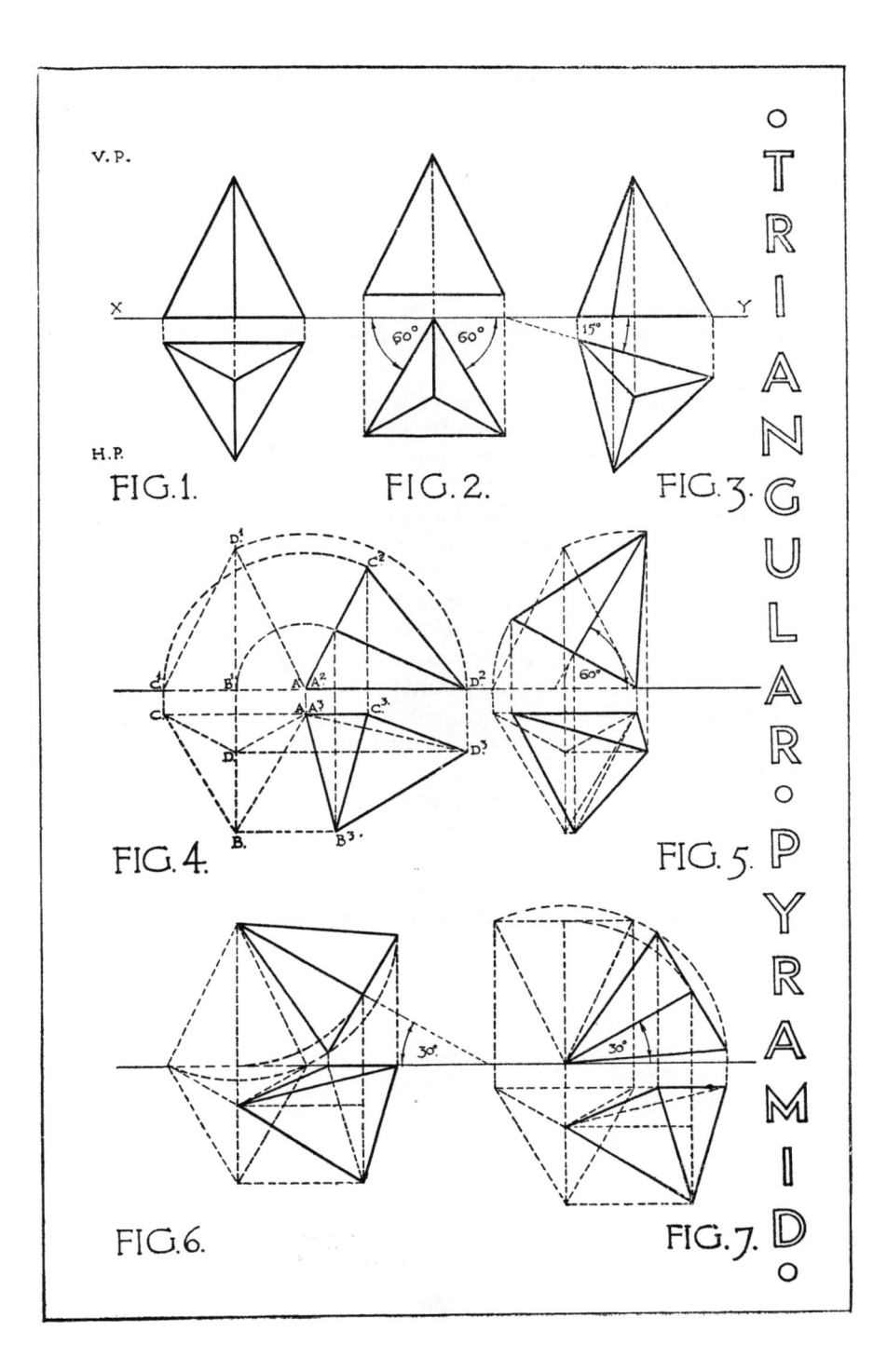

V.P.

X ——— Y

H.P.

FIG.1.   FIG.2.   FIG.3.

60° 60°   15°

FIG.4.   FIG.5.

60°

FIG.6.   FIG.7.

30°   30°

PROJECTIONS OF SOLIDS: HEXAGONAL PYRAMID

THE base of a hexagonal pyramid is a regular hexagon and each of the six sides of the solid is a triangle.

In Fig. 1 the pyramid is standing on H.P.; it is in front of V.P. with one base edge parallel to it.

The pyramid in Fig. 2 has its base parallel to and above H.P. One base angle is in contact with V.P. and two adjacent base edges are at 30° to V.P.

In Fig. 3 the solid is standing on H.P., it is in front of V.P. and one base edge is at an angle of 45° to V.P.

The example in Fig. 4 is resting with one corner on H.P., it is in front of V.P. with the axis parallel to it.

Fig. 5 shows a hexagonal pyramid in space with its base inverted. It has its apex above H.P. and its axis parallel to V.P. and at 30° to H.P.

In Fig. 6 the pyramid is lying with one of its triangular sides on H.P. One base angle is in contact with V.P. and the axis is parallel to V.P.

The pyramid in Fig. 7 is above H.P. with one base edge in contact with V.P., its axis is at 45° to V.P. and parallel to H.P.

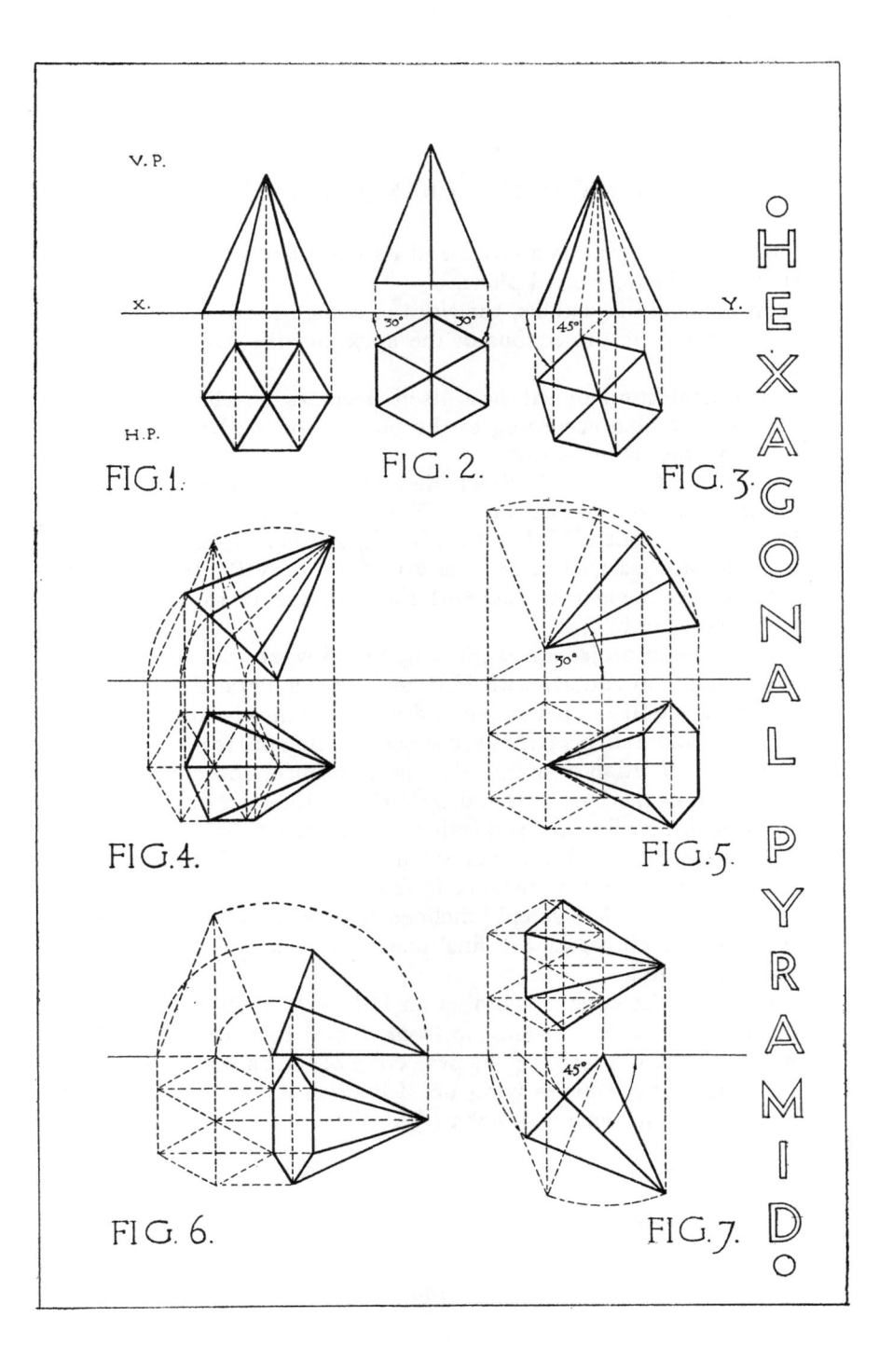

HEXAGONAL PYRAMID

V.P.

X. Y.

H.P.

FIG. 1.   FIG. 2.   FIG. 3.

30°   30°   45°

FIG. 4.   FIG. 5.

30°

FIG. 6.   FIG. 7.

45°

PROJECTIONS OF SOLIDS: CONE

The plan of a cone is a circle and its elevation a triangle when it stands on the horizontal plane.

In inclined projections, the sides of a cone remain straight lines drawn from the projections of the apex and the diameters of the base.

The projection of the base itself becomes an ellipse either in elevation or plan according to the position of the base in relation to the planes of projection.

In Figs. 1, 2, and 3 direct views are shown which are either triangles or circles in each case. The cone in Fig. 1 is standing on H.P. and in front of V.P., whilst in Fig. 2 it is in space above H.P. with its base parallel to H.P. and in front of V.P. In Fig. 3 the base of the cone is in contact with the V.P. with its axis above H.P. and parallel to it.

An inclined projection is given in Fig. 4 where the cone is lying with its side in contact with H.P. and its axis parallel to V.P.

The elevation and plan are first drawn with the elevation standing on H.P. and the plan in its correct position in relation to V.P. Turn the elevation over into its new position, lying on H.P. The projection of the base is found by dividing the plan into a number of parts and finding the projection of each point, as 7, 7^1, 7^2, 7^3.

The cone in Fig. 5 is standing with its apex on H.P. and its axis at 45° to H.P., a given distance in front of V.P. and parallel to it.

As the base is less steeply inclined to H.P. than in the previous example, the ellipse in the final plan more nearly approaches the shape of a circle.

In Fig. 6 the cone is standing on H.P. with its base at 45° to it and its axis a given distance in front of and parallel to V.P. Half only of the ellipse forming the projection of the base is seen in plan.

In Fig. 7 the cone is lying on H.P. with its axis parallel to it. The base of the cone is at 45° to V.P.

V.P.

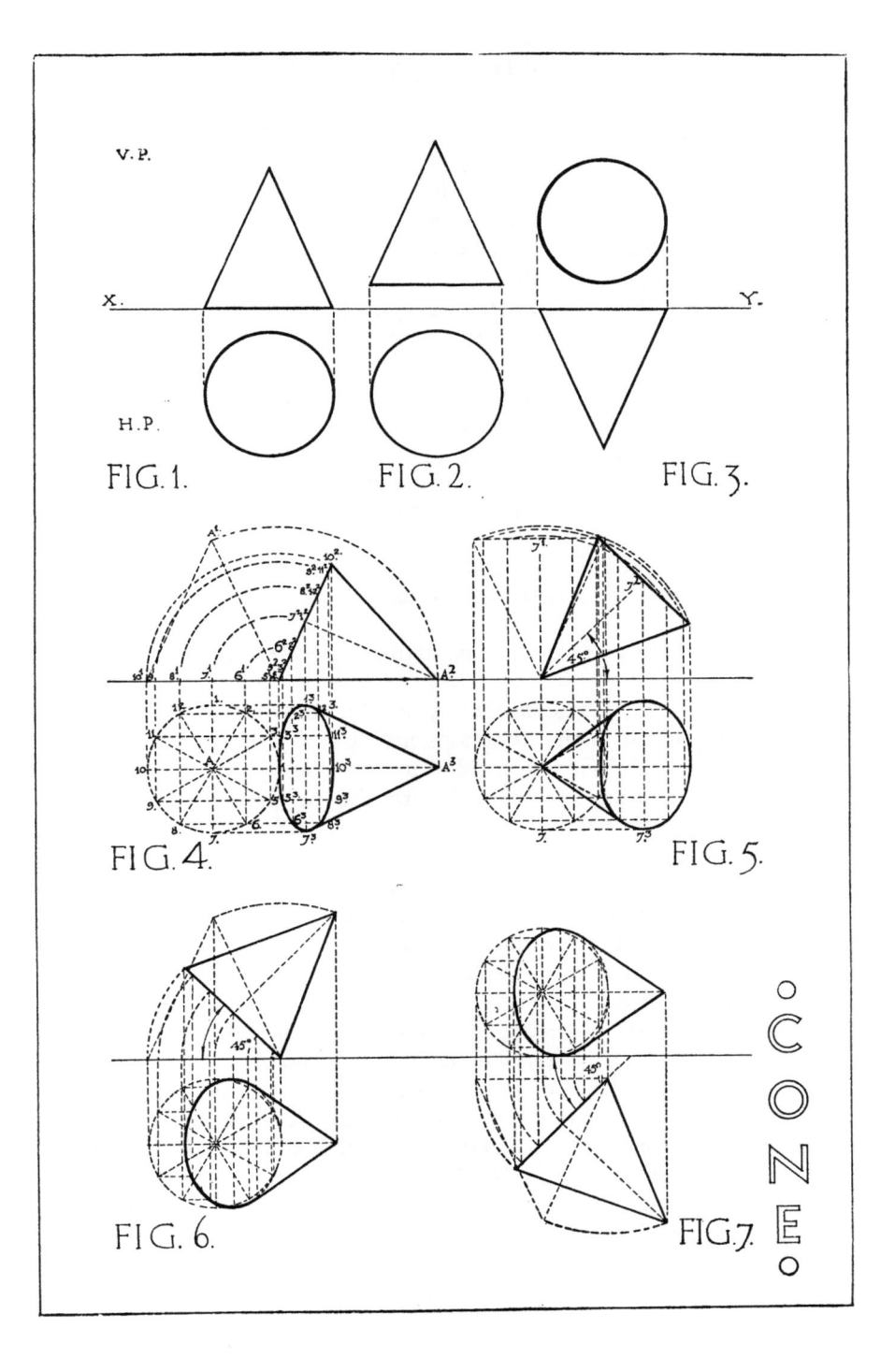

X. ——————————————————————————— Y.

H.P.

FIG. 1.   FIG. 2.   FIG. 3.

FIG. 4.   FIG. 5.

FIG. 6.   FIG. 7.

CONE

AUXILIARY PROJECTION

An auxiliary projection is one made on to an imaginary plane other than the horizontal, vertical, or side vertical planes. This subsidiary plane is imagined to be inserted at right angles to the direction of view. Normal elevations and plans are, of course, views at right angles to the horizontal and vertical planes respectively. Auxiliary projection is also known as *change of ground line*.

The procedure for drawing auxiliary ELEVATIONS is:

(a) Insert new XY or ground line at right angles to direction of view.

(b) Project points from the *plan* on to the new ground line.

(c) Measure heights from *original elevation* and transfer them to the new elevation.

The procedure for drawing auxiliary PLANS is:

(a) Insert new XY or ground line at right angles to direction of view.

(b) Project points from the *elevation* on to the new ground line.

(c) Measure lengths from the *original plan* and transfer them to the new plan.

It is more usual to require an auxiliary elevation than an auxiliary plan.

In Figs, 1 and 2 a hexagonal pyramid is shown with two auxiliary elevations taken in the directions indicated by the arrows. In the first example the view is parallel to two sides of the prism so the auxiliary elevation is identical with the projection on the H.P. The second direction of view is at 45° to H.P. and shows two of the angular edges in elevation. The height of the apex, H, is the only dimension to transfer in this example.

The auxiliary elevation and auxiliary plan of an angle are shown in Figs. 3 and 4. The thickness of the base of the angle, B, is a height to be transferred to the new elevation in addition to the over-all height A.

In Fig. 4 after projecting the points from the elevation the only dimension to be transferred is the length of the plan L.

An H-shaped solid is given in Fig. 5 together with an auxiliary elevation at an angle of 30° to H.P. The depth of the horizontal bar may be transferred as indicated at C and B or added to the height C.

Solids composed partly or wholly of curves, such as the cylinder, Fig. 6, necessitate dividing the curved portion into a series of points, projecting and measuring each point as with previous examples

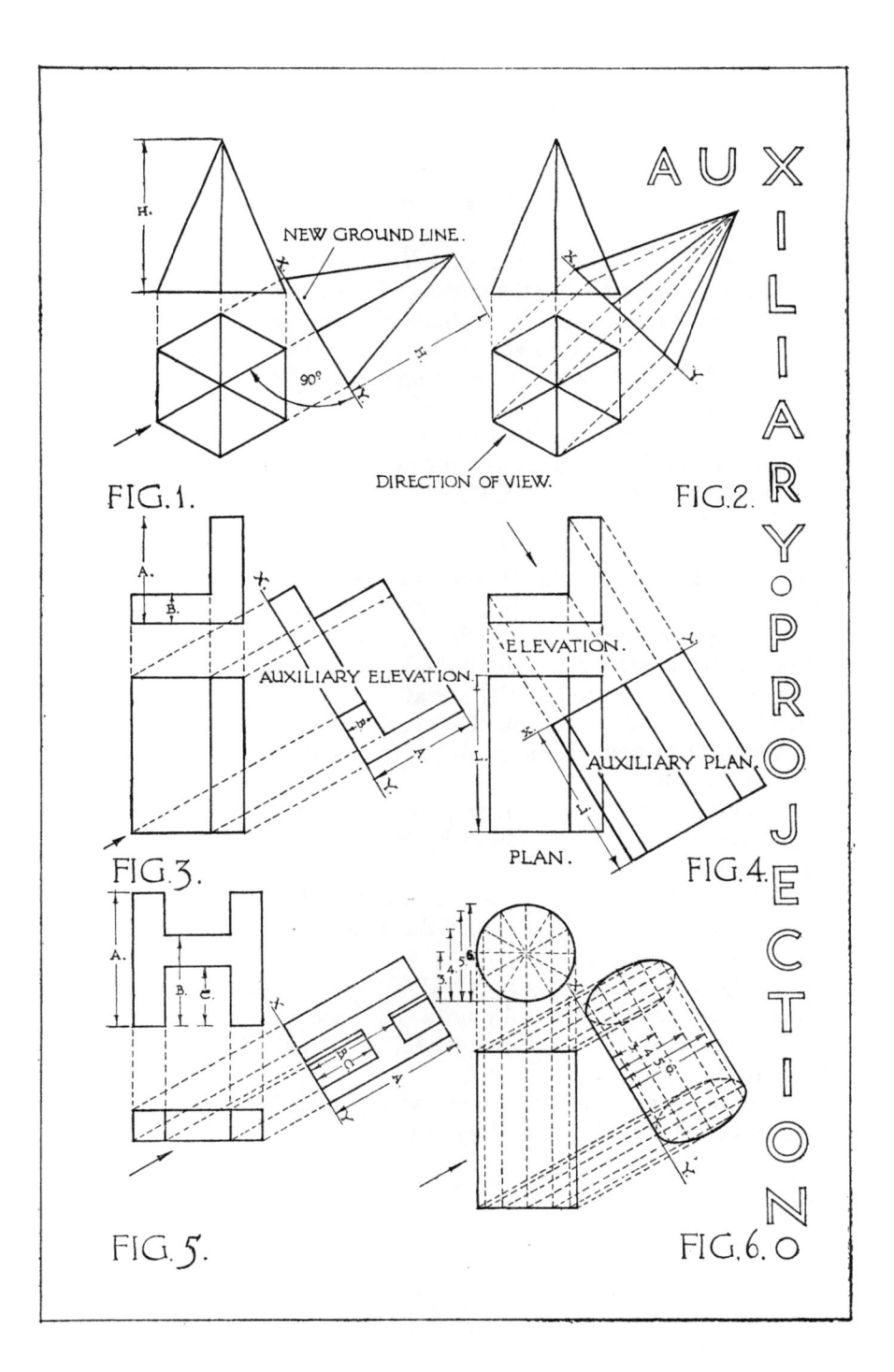

AUXILIARY PROJECTION

FIG. 1. NEW GROUND LINE. DIRECTION OF VIEW. 90° H.

FIG. 2.

FIG. 3. AUXILIARY ELEVATION.

FIG. 4. ELEVATION. AUXILIARY PLAN. PLAN.

FIG. 5.

FIG. 6.

SECTIONS OF SOLIDS

It is very often necessary in practice to determine the true shape of the section formed by the cutting of a solid by an inclined plane. There are several available methods of construction.

In Fig. 1 a square prism is shown standing with one angular edge on H.P. and its axis at right angles to V.P. The prism is cut by the plane CD.

It is required to project two auxiliary elevations of the solid from the direction indicated by the arrows and also to obtain the true shape of the cutting plane CD.

The two auxiliary elevations need no explanation, being constructed as previously explained, after inserting the new ground line at right angles to the direction of the view.

The cut section of the prism as seen in the auxiliary projection is not the true shape of the cutting. It would only be so if the direction of the arrow were at right angles to the plane CD. The true shape is an elongation of the elevation of the cutting, that is, the heights remain constant but the lengths increase.

Using D in plan as centre, rebat points F and C into the horizontal line DC^3.

Project vertically from these points until they meet their corresponding points carried across horizontally giving F^4, C^4, F^4, D^4, D^4 as the required section.

A hexagonal prism is given in Fig. 2, this time standing with one of its sides on H.P. It is cut by the plane AB and it is required to find the true shape of the cutting.

Using A as centre rebat points C, D, and B onto a horizontal line from A and project vertically until they meet horizontal projectors from the elevation of each point. A^2, C^2, D^2, E^2, E^2, D^2, C^2 is the required true shape.

A similar procedure is followed when using B as centre and the two sections are, of course, identical.

SECTIONS OF SOLIDS

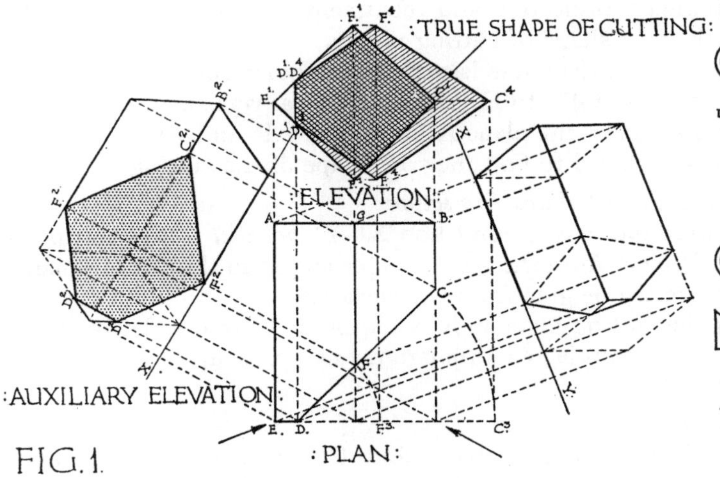

FIG.1.

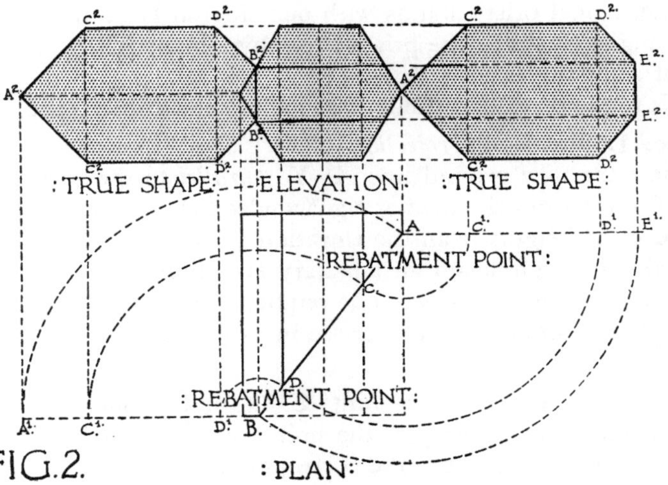

FIG.2.

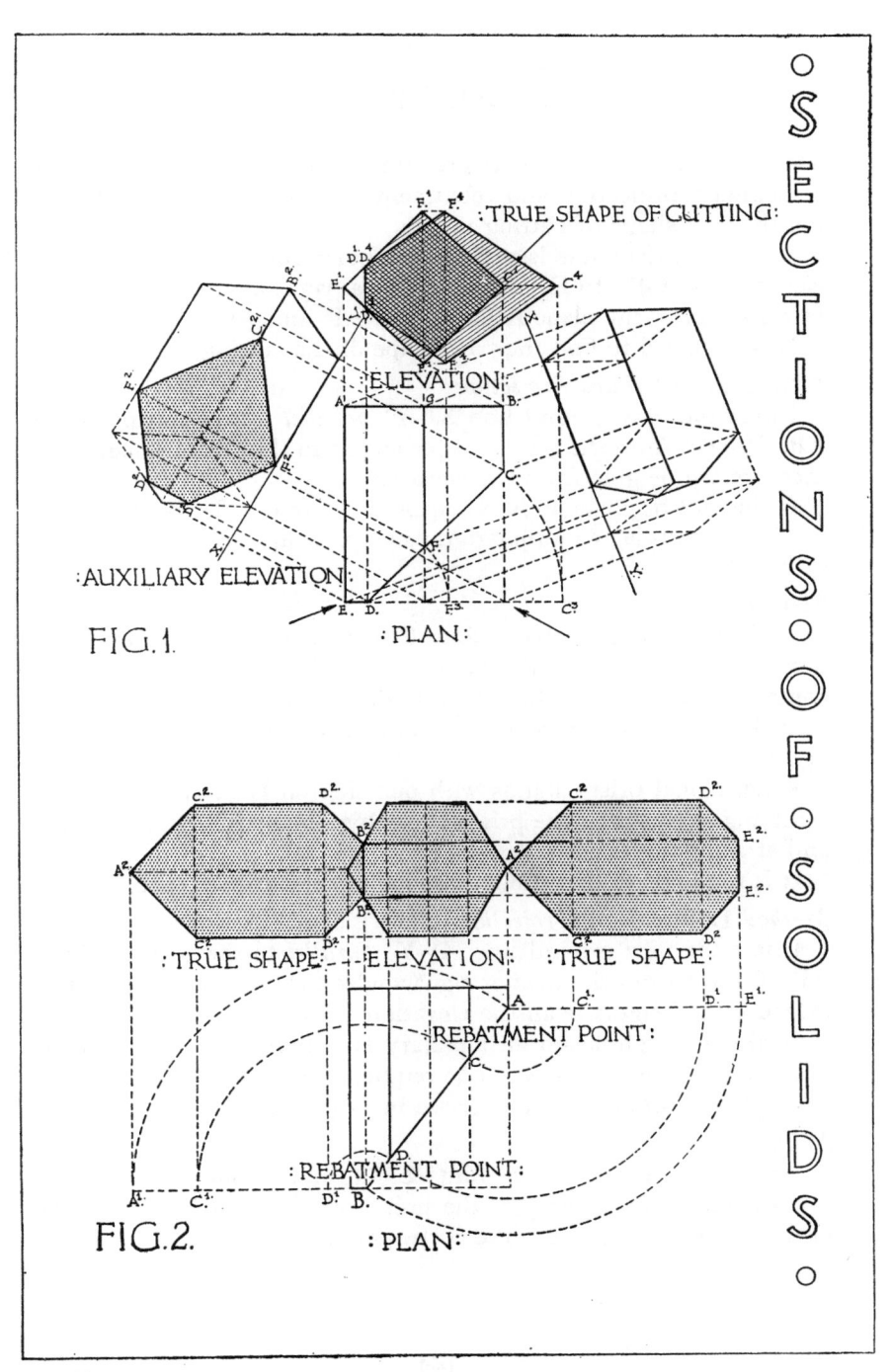

SECTIONS OF SOLIDS

THE two examples given here are intended to compare the methods of auxiliary projection and rebatment as constructions for obtaining the true shape of cutting planes.

A square prism stands with one angular edge on H.P., two adjacent sides at 45° to H.P. and its axis at right angles to V.P. The prism is cut by the plane AB at 60° to V.P. and 90° to H.P., Fig. 1.

It is required to find the true shape of the cutting.

Method 1. Auxiliary Projection

Insert the new ground line X^1–Y^1 *parallel* to the cutting plane AB. Project from the plan at right angles to the cutting plane and measure the heights from the elevation.

If the cutting plane only is required the remainder of the figure need not be projected although the complete figure has been shown in Fig. 1.

Method 2. Rebatment

With A as centre rebat points C and B onto a horizontal line drawn from A. Project the points vertically until they meet horizontal projectors from the corresponding points in elevation.

In Fig. 2 similar constructions have been used with a different figure.

A hexagonal prism stands with one side on H.P. and its axis at right angles to V.P. The prism is cut by a plane AB at 55° to V.P. and at right angles to H.P.

It is required to find the true shape of the cutting plane.

Method 1. Auxiliary Projection

Insert the new ground line X^1–Y^1 *parallel* to the cutting plane AB. Project from the plan at right angles to the cutting plane and measure the heights from the elevation.

In this example again an auxiliary elevation of the whole solid has been shown. This was not required but might very well be attempted as an additional exercise in projection.

Method 2. Rebatment

With A as centre, rebat points C, D, and B onto a horizontal line drawn from A. Project the points vertically until they meet horizontal projectors from the corresponding points in elevation.

SECTIONS OF SOLIDS

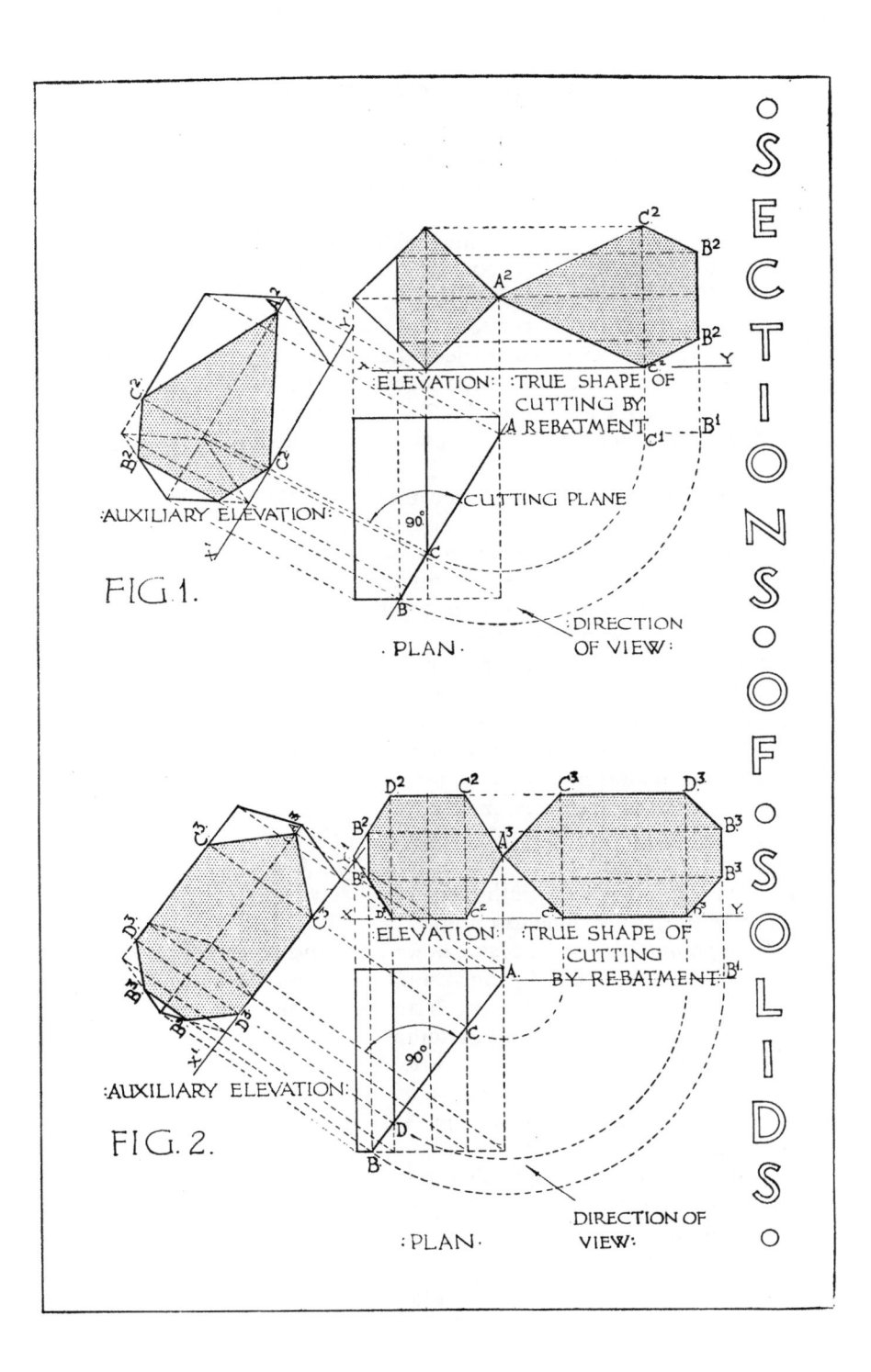

FIG. 1.

AUXILIARY ELEVATION
ELEVATION
TRUE SHAPE OF CUTTING BY A REBATMENT
CUTTING PLANE
PLAN
DIRECTION OF VIEW
90°

FIG. 2.

AUXILIARY ELEVATION
ELEVATION
TRUE SHAPE OF CUTTING BY REBATMENT
PLAN
DIRECTION OF VIEW
90°

SECTIONS OF SOLIDS

These examples show various cuttings of right pyramids, the projections of the plans of the cuttings and the true shapes of the cut surfaces.

In Fig. 1 a hexagonal pyramid is standing on H.P. with one base edge parallel to V.P. It is cut by a plane AB at 50° to H.P. and at 90° to V.P.; also by a plane CD at 53° to V.P. and 90° to H.P. It is required to find the true shape of the cutting planes.

From the elevation AB project the plan of the cutting by dropping down the points where the plane cuts the inclined edges of the pyramid giving 1', 2', 3', 4', 3', 2'.

The true shape may be obtained by projecting at right angles to the cutting, making the axis 1″–4″ parallel to it, and measuring widths from the plan. 2'–2' becomes 2″–2″, 3'–3' becomes 3″–3″, giving 1″, 2″, 3″, 4″, 3″, 2″ as the true shape.

If rebatment is used the lengths on the inclined plane are turned into the horizontal using A as centre. Point 1 could have been used as centre after drawing a horizontal from it, which would have made points 1' and 1″ coincide.

Cutting plane CD is inclined to V.P. but at 90° to H.P., therefore the plan of the cutting will be a straight line. The elevation is found by projecting vertically from the plan giving 5', 6', 7', 8'.

The true shape has been found by rebatment only using D as centre and taking horizontal projectors from the elevation; 5″, 8″, 7″, 6″ is the required section.

The square pyramid in Fig. 2 is cut by a plane AB at 45° to H.P. and at right angles to V.P., also by a plane CD at 45° to V.P. and at right angles to H.P. Both true shapes have been found by rebatment as in the previous example.

The points 2‴ on the plan cannot be found by direct projection. But 0–2‴ is the plan of a horizontal line from the axis to an edge of the pyramid, and since the latter is symmetrical we may find this distance by drawing a horizontal through 2 in the elevation to meet the original edge at 2'. Dropped on to the plan, this becomes 0–2″, and 0–2‴ is made equal to it.

The octagonal pyramid in Fig. 3 is cut by a plane AB at 45' to H.P. and vertical to V.P.

It is also cut by a plane CD at 90° to both H.P. and V.P. As CD is vertical to both planes it will appear as a straight line in elevation and in plan.

SECTIONS OF SOLIDS

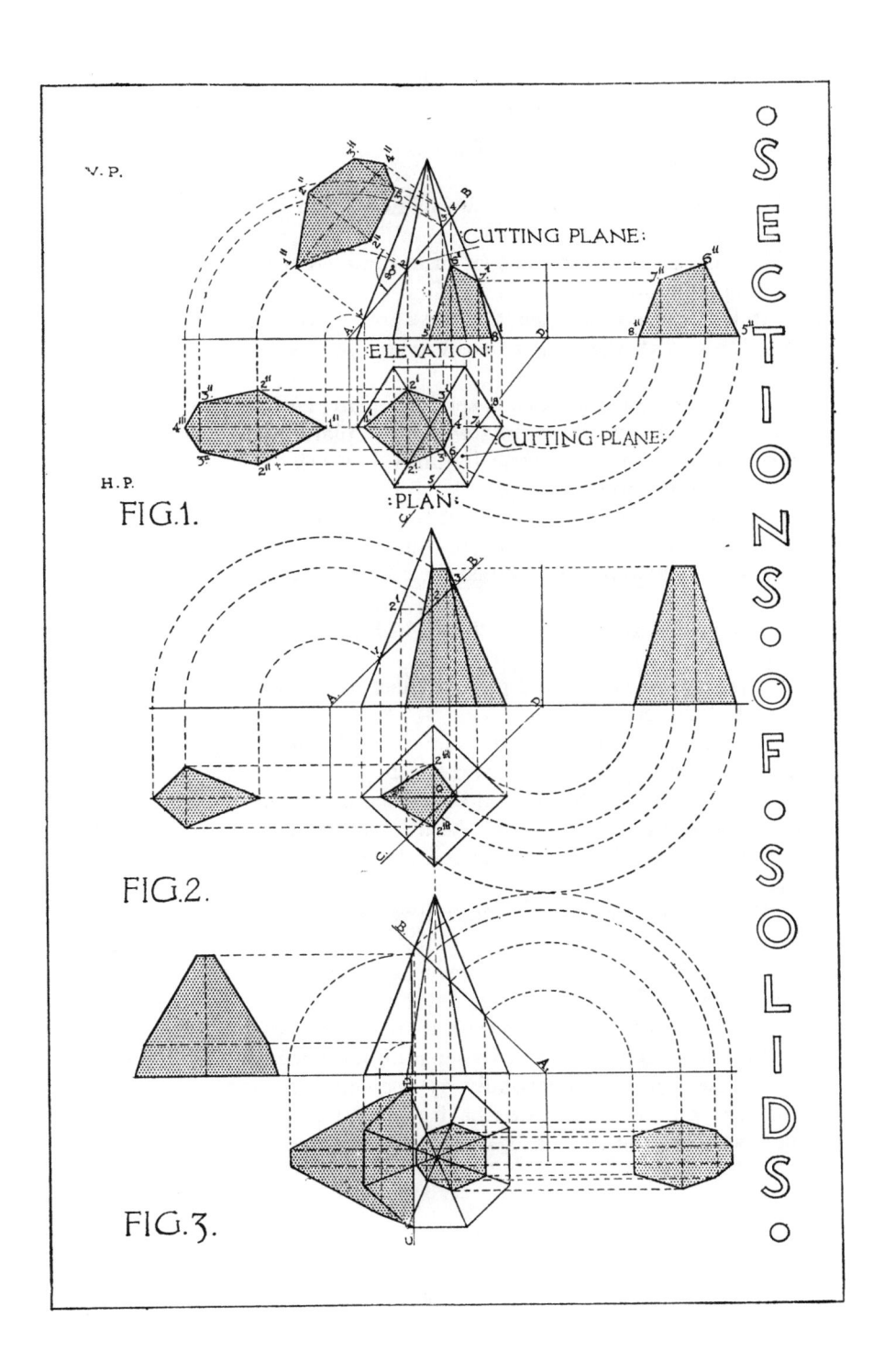

FIG. 1.

FIG. 2.

FIG. 3.

SECTIONS OF SOLIDS

THESE further examples show inclined cuttings through moulded sections.

In Fig. 1 the section is that of a moulded base section stopping on a splay stop.

It is required to draw the elevation of the intersection of the moulding with the stop and the true shape of the section at that point.

Draw the plan and normal section, taking additional points in the curve as at X. At the points where the plan of the moulding lines meet the stop, project vertically until they meet the elevation of the corresponding lines as at X^2.

The true shape can be found by rebatment. Transfer the true length of the stop and moulding lines to a horizontal line and project vertically until they meet their corresponding lines projected horizontally.

A moulded glazing bar is shown in Fig. 2 with a splay cut at one end.

It is required to draw a plan of the timber and to obtain the true shape of the cut surface.

As the cut is at an angle to the nose of the moulding the elevation and true section must be drawn first. Project the elevation of the moulding lines of the cut and project down until they meet the corresponding lines in plan. Additional points would be required in the curves, but for clearness have been omitted from this drawing.

The true shape has been found by rebatment as in the previous example.

A moulded jamb stone is shown in elevation and plan, Fig. 3, its top surface being cut at an angle of 45° to H.P.

It is required to find the true shape of the cutting.

This has been done by two separate methods, first by rebatment giving section A, and secondly by projecting at right angles to the cutting and measuring depths from plan giving section B. The elongation of the section has been in one direction only.

SECTIONS OF SOLIDS

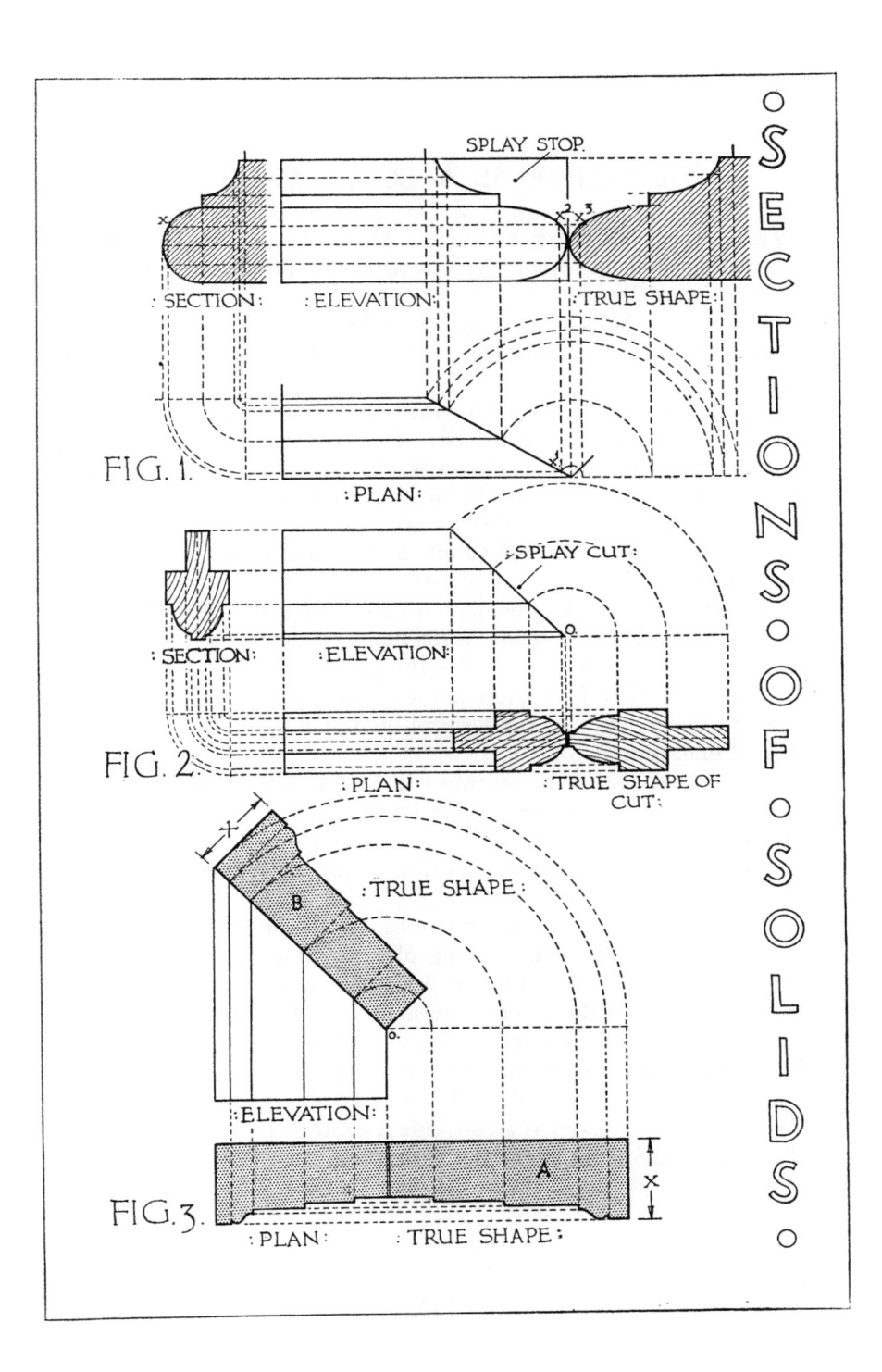

FIG. 1.
SPLAY STOP.
:SECTION: :ELEVATION: :TRUE SHAPE:
:PLAN:

FIG. 2.
SPLAY CUT:
:SECTION: :ELEVATION:
:PLAN: :TRUE SHAPE OF CUT:

FIG. 3.
:TRUE SHAPE:
:ELEVATION:
:PLAN: :TRUE SHAPE:

THE ELLIPSE AS A CONIC SECTION

If a cone or a cylinder is cut by a plane inclined to the axis and passing through the opposite sides of the solid the true shape of the cutting is an ellipse.

Various methods of drawing the ellipse as a plane figure have already been shown; here the curve is treated only as a section of a cone.

Let X–Y be the cutting plane, the true shape of which is required.

Divide the plan into a number of equal parts as at 1, 2, 3, etc., and connect the points to the apex O. These lines are generators of the cone and should also be drawn in elevation.

Find the plan of the cutting by taking the points where the plane cuts the generators in elevation and dropping them on to the generators in plan.

The plan of point $1'''$, which is on the vertical generator, is found by projecting $1''$ in elevation horizontally to the side of the cone as at X. The distance $1''$–X is also the distance $1''$ is in front of the vertical axis, therefore drop point X into plan and using O as centre transfer O–X to the vertical, giving $1'''$ and $1'''$ as the required plan points.

The true shape of the cutting will be an elongation of the plan in one direction only, the lengths having increased but the widths remaining the same.

Two methods of drawing the true shape are shown, direct projection and rebatment.

In direct projection a new centre line or major axis of the ellipse is inserted parallel to the cutting plane in elevation. The points where the generators intersect the plane are projected at right angles and the widths on each line measured from plan to obtain a series of points in the curve. The ellipse could have been drawn by any other method having determined the major and minor axes.

A similar result is obtained by rebatment, and in the example, Fig. 1, point C is taken as a convenient centre which gives the base of the cone produced as the horizontal line.

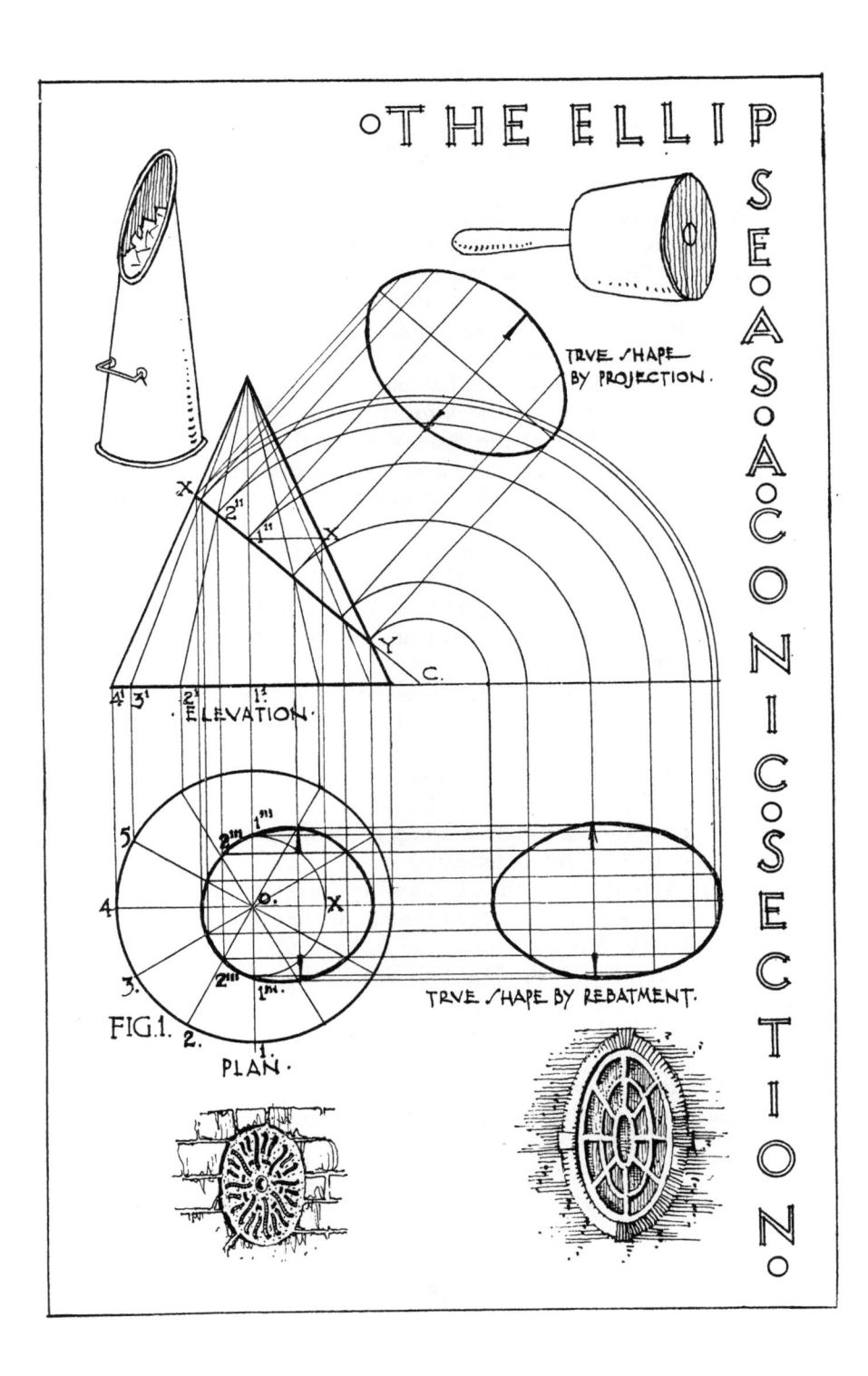

THE ELLIPSE AS A CONIC SECTION

TRUE SHAPE BY PROJECTION.

ELEVATION

TRUE SHAPE BY REBATMENT.

FIG.1.

PLAN.

PROJECTION OF POINTS

It is debatable if the study of the projection of points, lines and planes should be made before or after the projection of solids. A line in space inclined to two imaginary planes is something rather abstract and difficult for a student to visualize and fully appreciate. Theoretically, solids should follow planes, but it is usually easier if a student deals with at least a few solid examples first.

The position of a point can be determined by relating it to three planes of projection: (*a*) the Vertical Plane (V.P.); (*b*) the Horizontal Plane (H.P.); (*c*) the Side Vertical Plane (S.V.P.). These planes are shown in Figs. 1 and 2 with the angles between the planes numbered 1 to 4 in a clockwise direction. A point, line or solid may be situated in any of these planes.

Examples of a point in space are shown in Figs. 3 to 6; in each case a pictorial view and an orthographic view is given. The actual planes are shown and marked H.P. or V.P., but this is not necessary in problems. The X–Y or hinge line between the planes must of course always be shown.

A consistent method of lettering points is desirable, and it is usual to letter the actual point with a capital, thus A, the plan of the point with a small letter, thus *a*, and the elevation thus *a'*. The point A in Fig. 3 is situated above the horizontal plane and in front of the vertical plane, it is in the 1st angle. It should be noted that the side vertical plane has been ignored, the other planes being considered to be of indefinite extent.

In Fig. 4, A is below the H.P. and in front of V.P., it is in the 2nd angle. The planes are revolved about the X–Y in a clockwise direction so *a* and *a'* coincide.

In Fig. 5 A is below the H.P. and behind the V.P., it is in the 3rd angle.

In Fig. 6 A is above the H.P. and behind the V.P., it is in the 4th angle.

Simple models made of cardboard or stiff paper are easy to make and assist considerably in visualizing the planes.

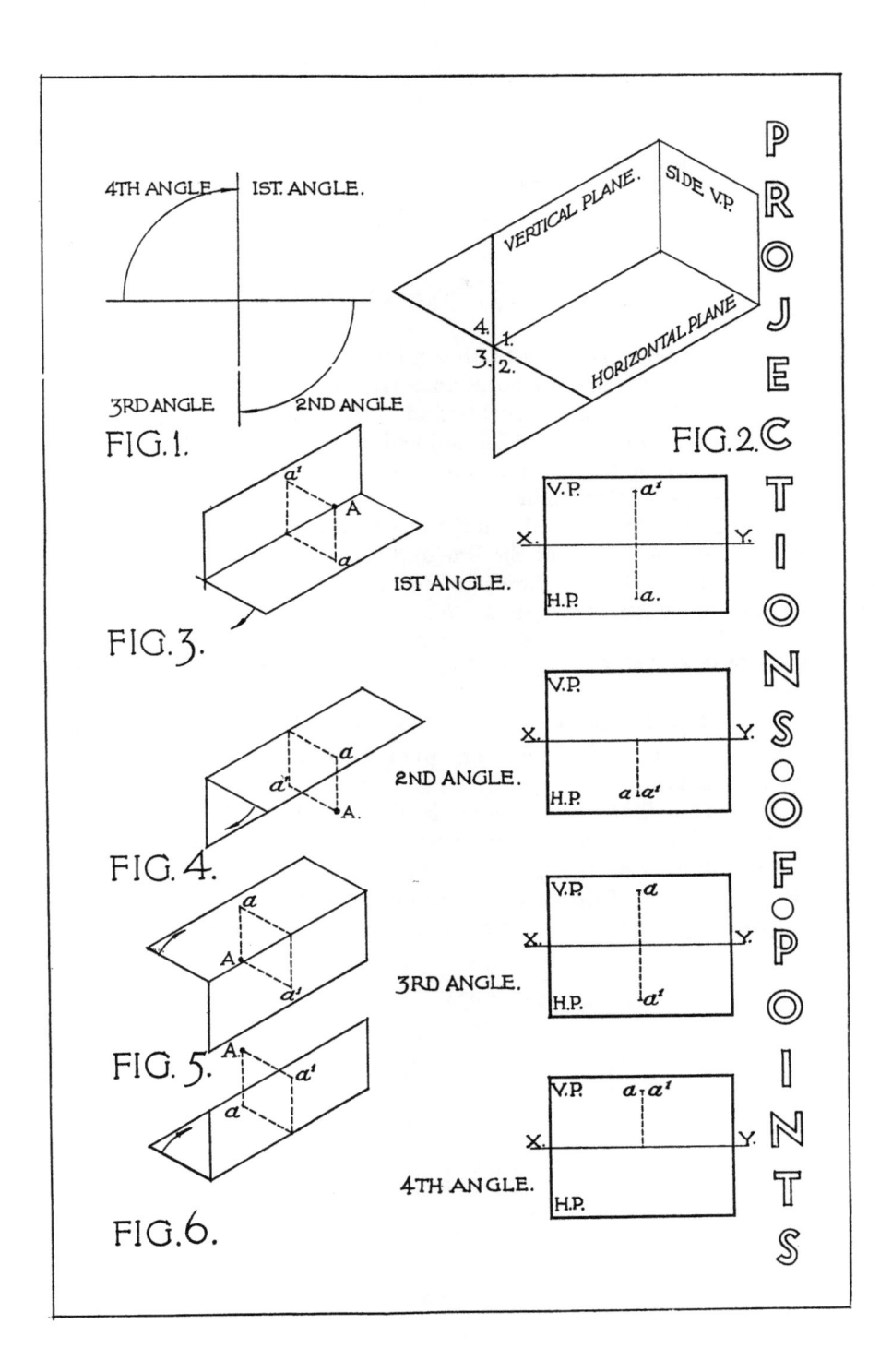

PROJECTIONS OF POINTS

FIG. 1.
4TH ANGLE | 1ST ANGLE
3RD ANGLE | 2ND ANGLE

FIG. 2.
VERTICAL PLANE. SIDE V.P.
HORIZONTAL PLANE
4. 1.
3. 2.

FIG. 3. 1ST ANGLE.
V.P. a'
X — Y
H.P. a

FIG. 4. 2ND ANGLE.
V.P.
X — Y
H.P. $a \; a'$

FIG. 5. 3RD ANGLE.
V.P. a
X — Y
H.P. a'

FIG. 6. 4TH ANGLE.
V.P. a , a'
X — Y
H.P.

PROJECTIONS OF LINES

A LINE can be considered to be terminated by two points.

A line in space may be perpendicular, horizontal, inclined to one or other of the planes, or inclined to both planes, i.e. oblique.

In Figs. 1 to 6 the pictorial and orthographic representation of a line A–B in each of these positions is shown. The line in all cases is in the 1st projection angle.

In Fig. 1 AB is vertical and is in front of the V.P. The elevation $a'b'$ is the true length of the line and the plan is the one point ab.

A horizontal line is shown in Fig. 2, its true length is the plan ab and its elevation the point $a'b'$.

Another horizontal line is shown in Fig. 3, AB being parallel to both H.P. and V.P. The true length of the line is in both elevation and plan.

AB in Fig. 4 is a line parallel to V.P. but inclined to H.P., its true length is the elevation $a'b'$. The plan of the line is ab but it is not its true length.

Another inclined line is given in Fig. 5. Here AB is parallel to H.P. and inclined to V.P., the true length is now the plan line ab.

An oblique line is represented in Fig. 6, AB being inclined to both the planes of projection $a'b'$ and ab are the elevation and plan respectively of the line, but neither represents its true length.

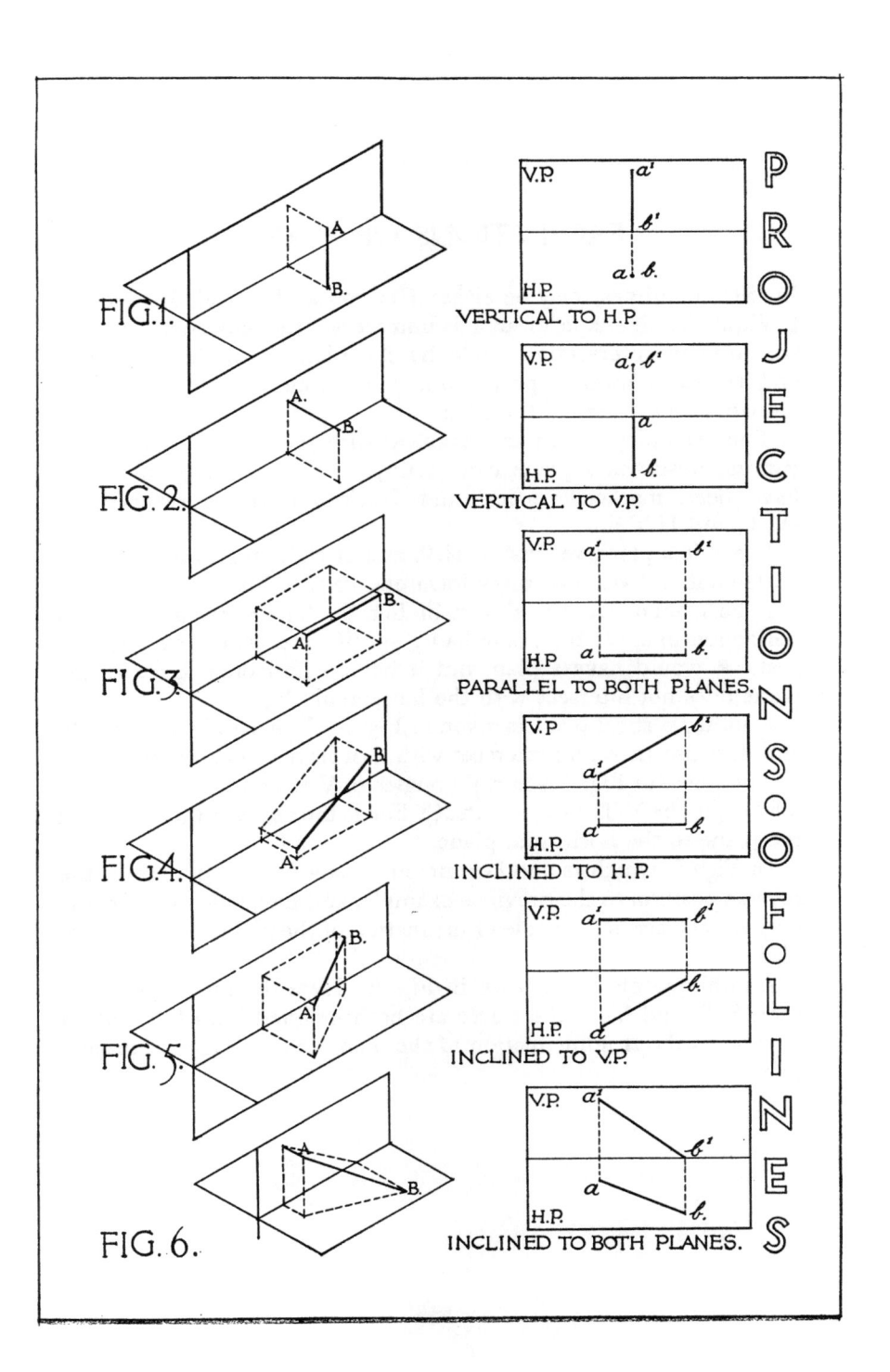

PROJECTIONS OF LINES

FIG. 1. VERTICAL TO H.P.

FIG. 2. VERTICAL TO V.P.

FIG. 3. PARALLEL TO BOTH PLANES.

FIG. 4. INCLINED TO H.P.

FIG. 5. INCLINED TO V.P.

FIG. 6. INCLINED TO BOTH PLANES.

PROJECTIONS OF PLANES

PLANES, like lines, may be either Horizontal, Vertical, Inclined or Oblique. Their shape, or area, is immaterial, the only concern being the lines of intersection made by the plane with the horizontal and vertical planes of projection. These lines of intersection are known as the *traces* of the plane.

The examples shown are given, as with the line, to indicate the varying positions a plane can occupy. The elevations and plans have been marked Vertical and Horizontal traces respectively (V.T. and H.T.).

Fig. 1 is a plane vertical to H.P. and at right angles to V.P., the horizontal and vertical traces form one vertical line.

A horizontal plane is shown in Fig. 2. The vertical trace is a horizontal line. If the plane had a specific size as the sketch suggests, it would have a plan, but it has no horizontal trace as the plane does not intersect with the horizontal plane.

Another vertical plane is given in Fig. 3. It is parallel to the V.P. and does not therefore intersect with it or have a vertical trace.

The plane in Fig. 4 is at right angles to V.P. and inclined to H.P. The angle the V.T. makes with XY line is the angle of inclination of the plane to the horizontal plane.

In Fig. 5, which is another inclined plane, the traces are in the reverse position to the previous example, and the angle made by the horizontal trace is the angle of inclination of the plane to the vertical plane.

An oblique plane is shown in Fig. 6. Here the inclination is to both H.P. and V.P., the traces are both inclined lines but neither represents the true inclination of the plane to the planes of projection.

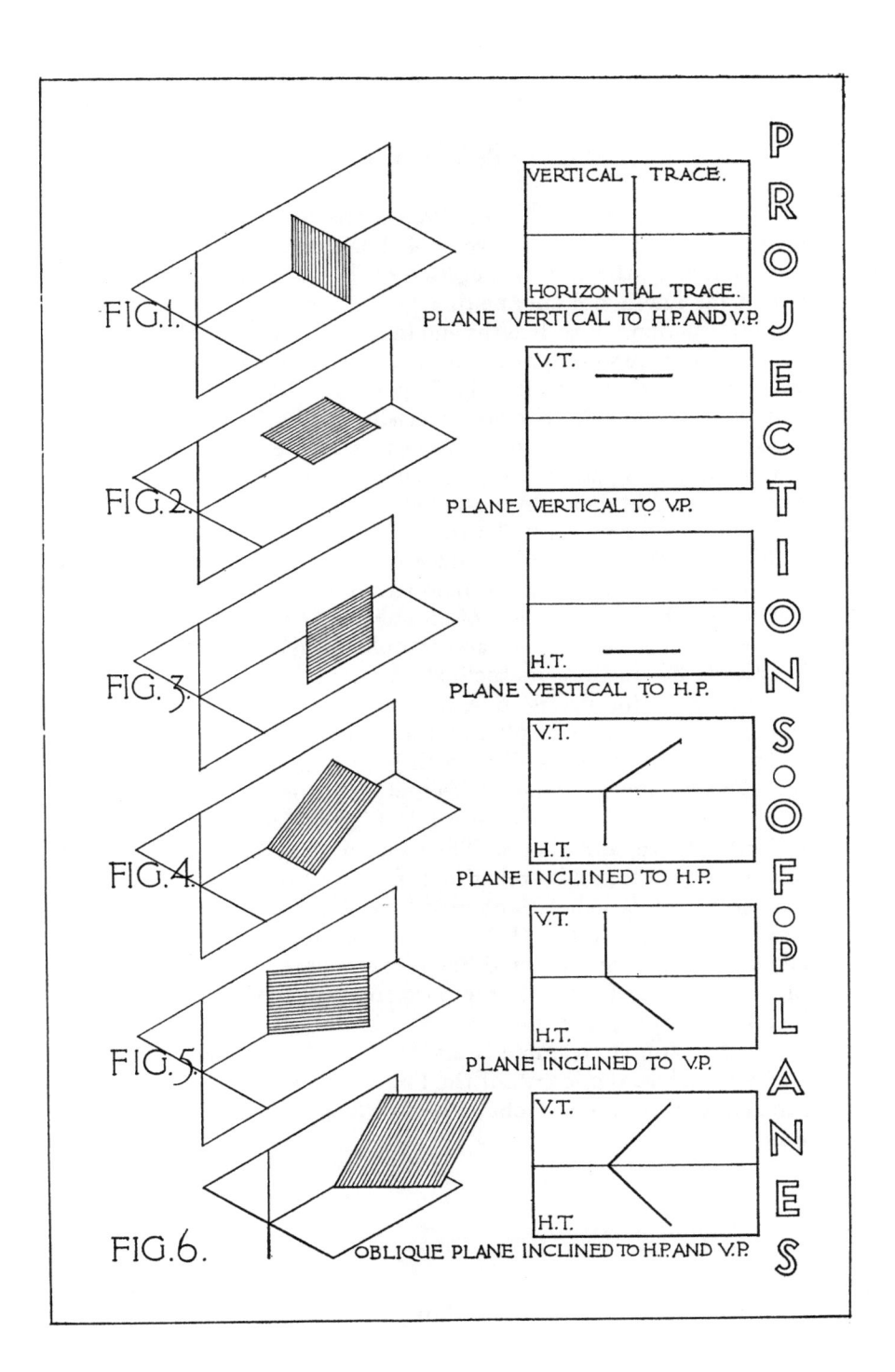

FIG.1.	VERTICAL TRACE / HORIZONTAL TRACE
	PLANE VERTICAL TO H.P. AND V.P.
FIG.2.	V.T.
	PLANE VERTICAL TO V.P.
FIG.3.	H.T.
	PLANE VERTICAL TO H.P.
FIG.4.	V.T. / H.T.
	PLANE INCLINED TO H.P.
FIG.5.	V.T. / H.T.
	PLANE INCLINED TO V.P.
FIG.6.	V.T. / H.T.
	OBLIQUE PLANE INCLINED TO H.P. AND V.P.

PROJECTIONS OF PLANES

TRUE LENGTHS OF LINES

It was seen in the section on Projections of Lines that the true length and inclination of a vertical, horizontal or inclined line is given either on the plan or elevation of the line. It was shown also that if the line is oblique, neither the plan nor the elevation gives this information. True lengths and inclinations may be found either by auxiliary views or by rebatment, and examples of both methods are shown in Figs. 1 to 8. Fig. 2 gives the elevation $a'b'$ and the plan ab of the line AB shown pictorially in Fig. 1.

An auxiliary view has been taken to find the true length of the line by projecting at right angles from the elevation and measuring from plan. The answer has also been obtained by projecting from plan and measuring from elevation.

The same line has been considered in Fig. 3, and its true length found twice, by rebatment in plan and in elevation.

Fig. 4 is the pictorial view of an oblique line AB. The elevation of the line is at 30° to H.P. and the plan at 45° to V.P. The true length of the line and its inclination to the V.P. are found by an auxiliary elevation in Fig. 6, and its true length and inclination to the H.P. by an auxiliary plan on the same drawing.

The same problem is dealt with in Fig. 5 and the true length found by rebatment in both plan and elevation.

The diagonal of a cube is shown in Fig. 7. The traces of the line will be at 45° to both planes. The true length and inclination of the line have been found by rebating the vertical trace into plan. The answer could have been obtained as easily by an auxiliary view, either elevation or plan. If the cube is considered to be cut diagonally across, then the required line is the hypotenuse or a triangular plane, the base of which is the horizontal trace of the line, and its height is the height of the cube.

Another simple example is afforded by the angular edges of a pyramid. Here the length of the line on plan, the H.T., has been rebated into elevation, to obtain its true length and angle of inclination, Fig. 8.

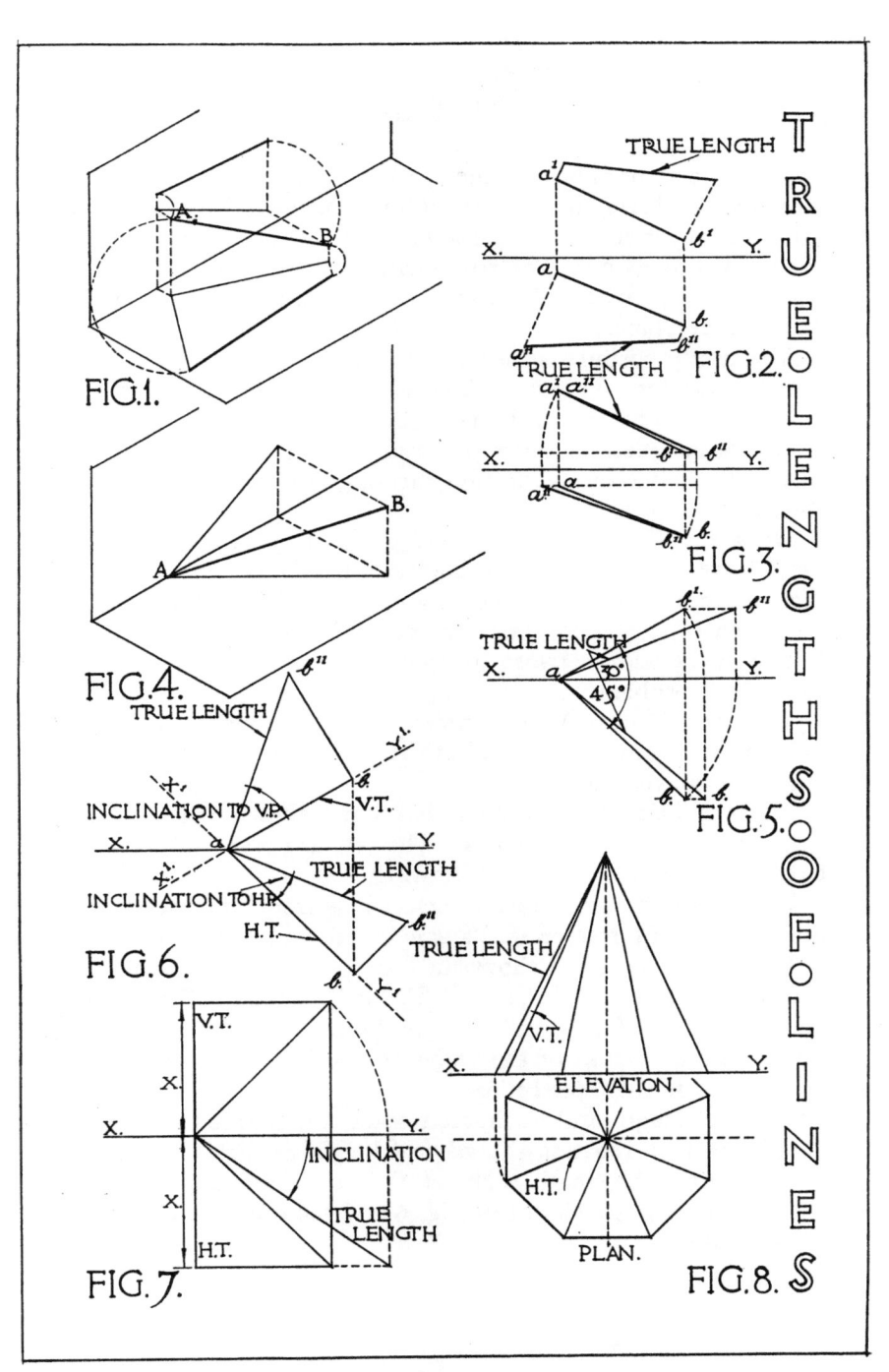

TRUE LENGTHS OF LINES

FIG. 1.

FIG. 2. TRUE LENGTH

FIG. 3. TRUE LENGTH

FIG. 4. TRUE LENGTH

FIG. 5. TRUE LENGTH

FIG. 6. INCLINATION TO V.P. / INCLINATION TO H.P. / TRUE LENGTH / V.T. / H.T.

FIG. 7. V.T. / H.T. / INCLINATION / TRUE LENGTH

FIG. 8. V.T. / H.T. / ELEVATION / PLAN

L

THE OBLIQUE PLANE

An oblique plane is one inclined to both planes of projection. The principal problem involved in oblique planes is to find the true angle of inclination of the plane to either the horizontal or vertical planes of projection. The true inclination is the angle made with the plane of projection by a cutting at right angles to the trace of the oblique plane.

The first procedure is therefore to insert a new plane at any convenient position. This new plane will be an inclined plane, therefore one of its traces will be vertical.

An oblique plane is shown pictorially in Fig. 1 with the new plane inserted. It is required to find the angle of inclination the plane makes with the H.P.

Draw the traces of the oblique plane (Fig. 2) and also of the newly inserted plane at right angles to the H.T. Erect a perpendicular from C and rebat the height of the V.T. (CD) to B. Join BA when BAC is the required angle. BA is the true length of the line of intersection between the two planes, but this does not form part of the problem.

The angle of inclination between the oblique plane and the V.P. is found in a similar manner, but the new plane is inserted at right angles to the vertical trace.

For simplicity this problem has been shown with the same oblique plane in Figs. 3 and 4. The two answers could very well have been found on the same drawing.

When two oblique planes intersect it is frequently necessary to find the line of intersection, its true length and inclination.

Two such planes are shown in Fig. 5. There will be traces to both planes, V.T.[1], V.T.[2], and H.T.[1], H.T.[2], Fig. 6. The traces of the intersection, V.T.[3] and H.T.[3], are found by projecting vertically from the intersection of the traces to the X–Y and joining these points back to the intersections.

The true length and inclination of the intersection line may be found either by rebating the vertical height of the V.T. into plan or by rebating the length of the H.T. into elevation.

Both methods are shown in Fig. 6 and the same result obtained in each case.

THE OBLIQUE PLANE

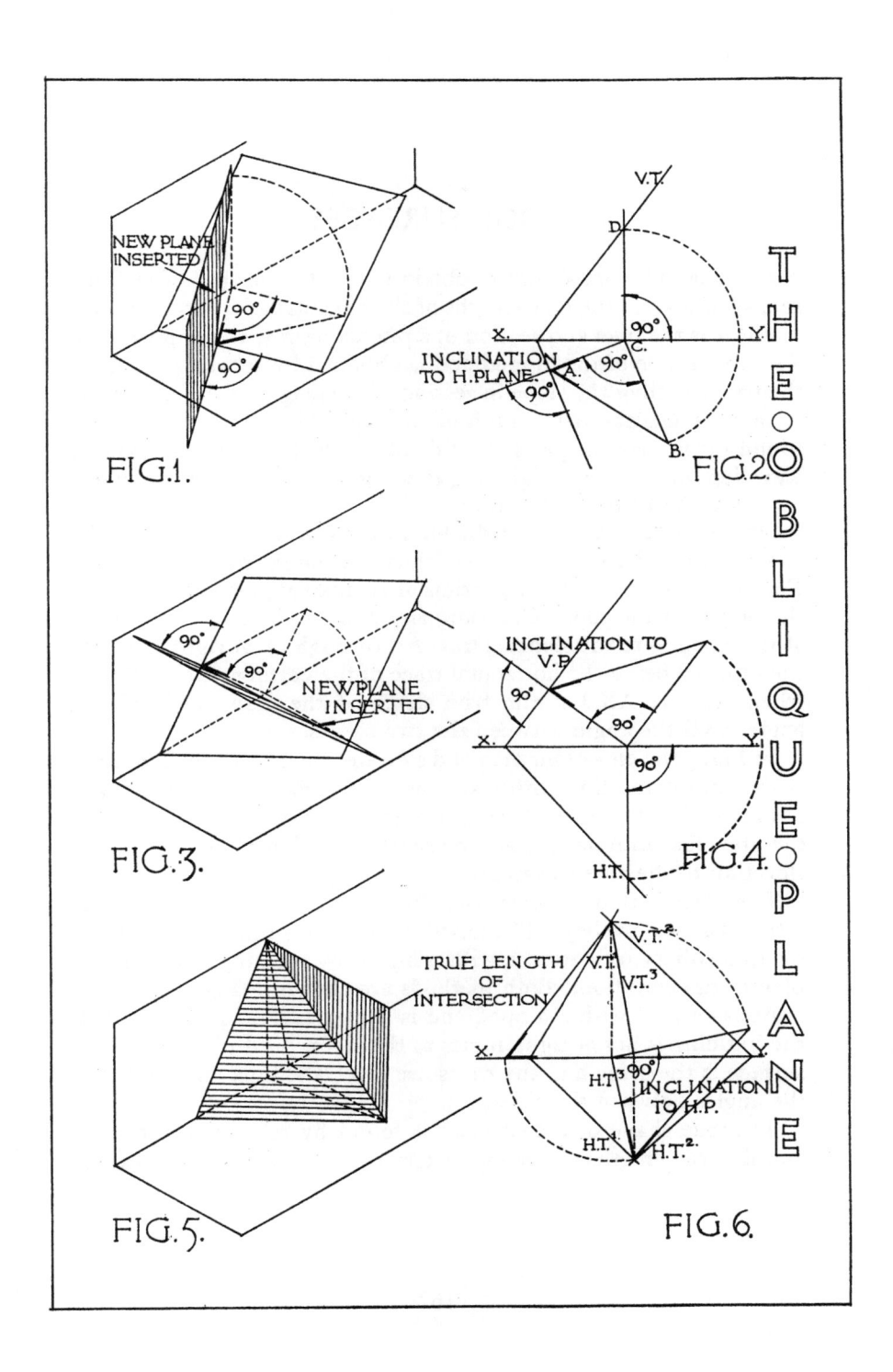

FIG. 1. NEW PLANE INSERTED. 90° 90°

FIG. 2. V.T. D INCLINATION TO H.PLANE. 90° 90° 90° A C B X Y

FIG. 3. 90° 90° NEW PLANE INSERTED.

FIG. 4. INCLINATION TO V.P. 90° 90° 90° X Y H.T.

FIG. 5. TRUE LENGTH OF INTERSECTION

FIG. 6. V.T.² V.T. V.T.³ X Y H.T.³ 90° INCLINATION TO H.P. H.T.⁴ H.T.²

ROOF SURFACES

THREE examples are shown of obtaining the true shapes of inclined roof surfaces and the true lengths of the hip and valley intersections.

Fig. 1 is the plan and section of a pitched roof with a hipped end. The true shape of the hipped end is obtained by rebating into plan the true length of the slope in section. The length of the side of the triangle is also the true length of the hip. The true length is also found in the true shape of the side of the roof and by an auxiliary elevation, projecting at right angles to the plan of the hip and marking on h, the height of the roof.

Obtaining the backing for the hip rafter, or the angle between the roof slopes, is shown in Fig. 2, which is an enlarged portion of Fig. 1. Find the slope of the hip by means of an auxiliary elevation using h, the height of the roof. Choosing any point P, insert a cutting PQ at right angles to the slope. Draw AB through Q, at right angles to the hip, AB being the horizontal trace of the cutting. Rebat P to X when triangle AXB is the true shape of the plane PQ and the angle AXB the angle between the two slopes.

In Fig. 3 the elevation, plan and side elevation of a gabled dormer is shown, intersecting with a sloping roof surface. The true shape of one of the sloping surfaces is found by rebating into plan, as in the previous example, and also by rebating the length of the slope in elevation to the side elevation.

The intersection between the dormer and the main roof is an internal angle, a valley. The length of the valley rafter can be found by an auxiliary elevation as with a hip or by rebating the length in plan to the elevation. Both methods are shown in Fig. 2.

Another roof with a hipped end is shown in Fig. 4. The end of the building is not at right angles to the sides, but the slope of each surface is the same and the eaves are level, the hips on plan bisect the angles between the sides.

The true shape of the surfaces is found by rebatment into plan and the true lengths of the hip rafters by rebatment into elevation.

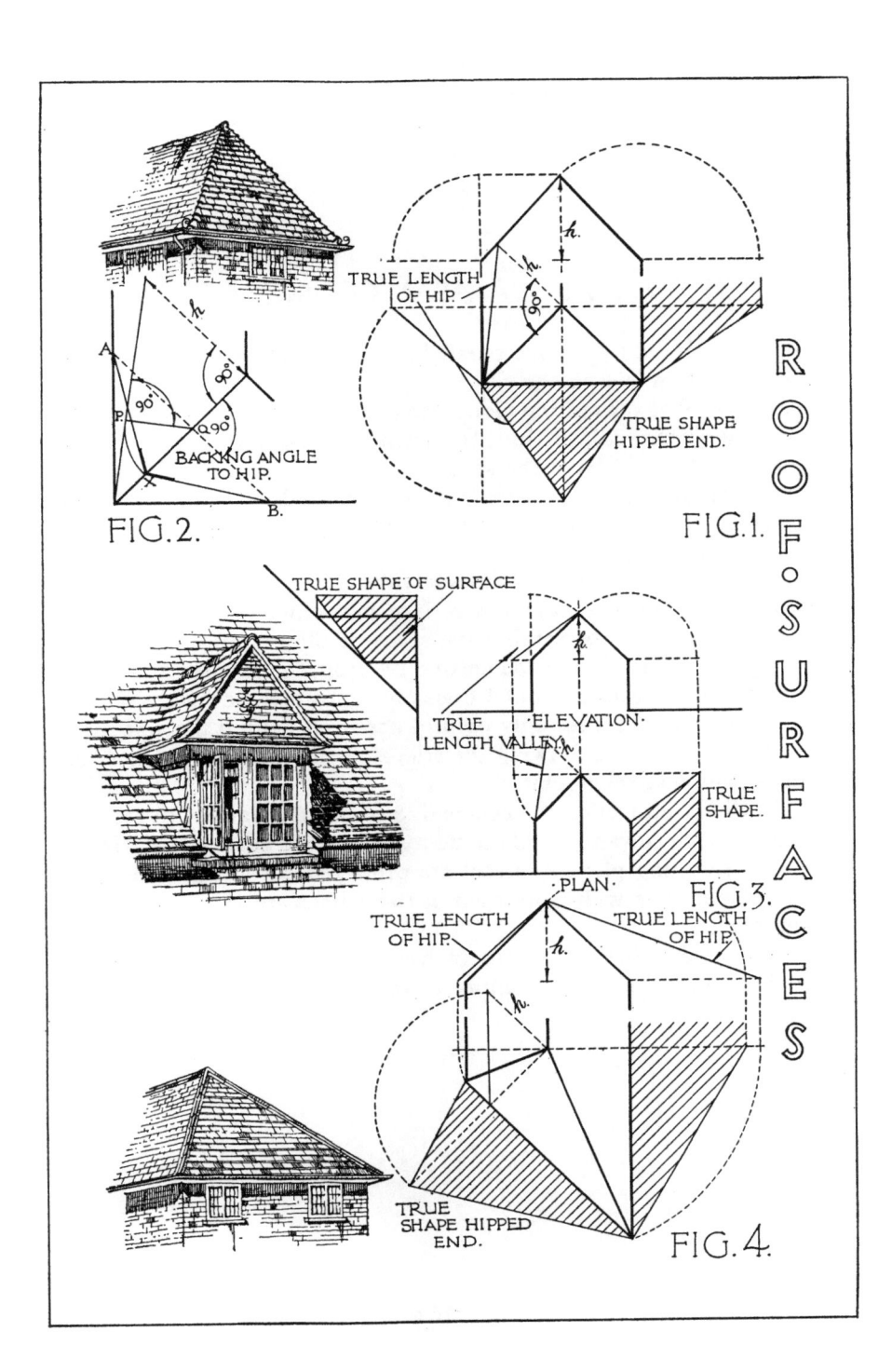

ROOF SURFACES

FIG. 1. TRUE LENGTH OF HIP. TRUE SHAPE HIPPED END.

FIG. 2. BACKING ANGLE TO HIP.

FIG. 3. TRUE SHAPE OF SURFACE. TRUE LENGTH VALLEY. TRUE ELEVATION. TRUE SHAPE. PLAN.

FIG. 4. TRUE LENGTH OF HIP. TRUE LENGTH OF HIP. TRUE SHAPE HIPPED END.

THE PARABOLA AS A CONIC SECTION

If a cone is cut by a plane parallel to one of its sides the true shape of the section is a parabola.

Let X–Y be the cutting plane, the true shape of which is required.

As with the ellipse the plan must first be drawn in order to ascertain the widths of the section.

Generators of the cone could have been used in the construction, but it is usually more convenient in this problem to use a series of horizontal divisions.

By means of horizontal lines divide either X–Y or the vertical height between the base and X into a number of parts. Draw a series of circles representing the plans of these lines.

Complete the plan of the cutting by dropping points from where the plane cuts the horizontal divisions on to their respective circles in plan. For clearness only point 4 has been figured throughout.

Two methods of drawing the true shape have been given: direct projection and rebatment.

In direct projection the centre line is inserted parallel to the cutting plane in elevation, and the points where the horizontal divisions and the cutting plane intersect are projected out at right angles to the plane. The widths on each of these lines are transferred from the plan.

The true shape by rebatment has been found as previously explained, using Y, the extremity of the cutting plane, as centre.

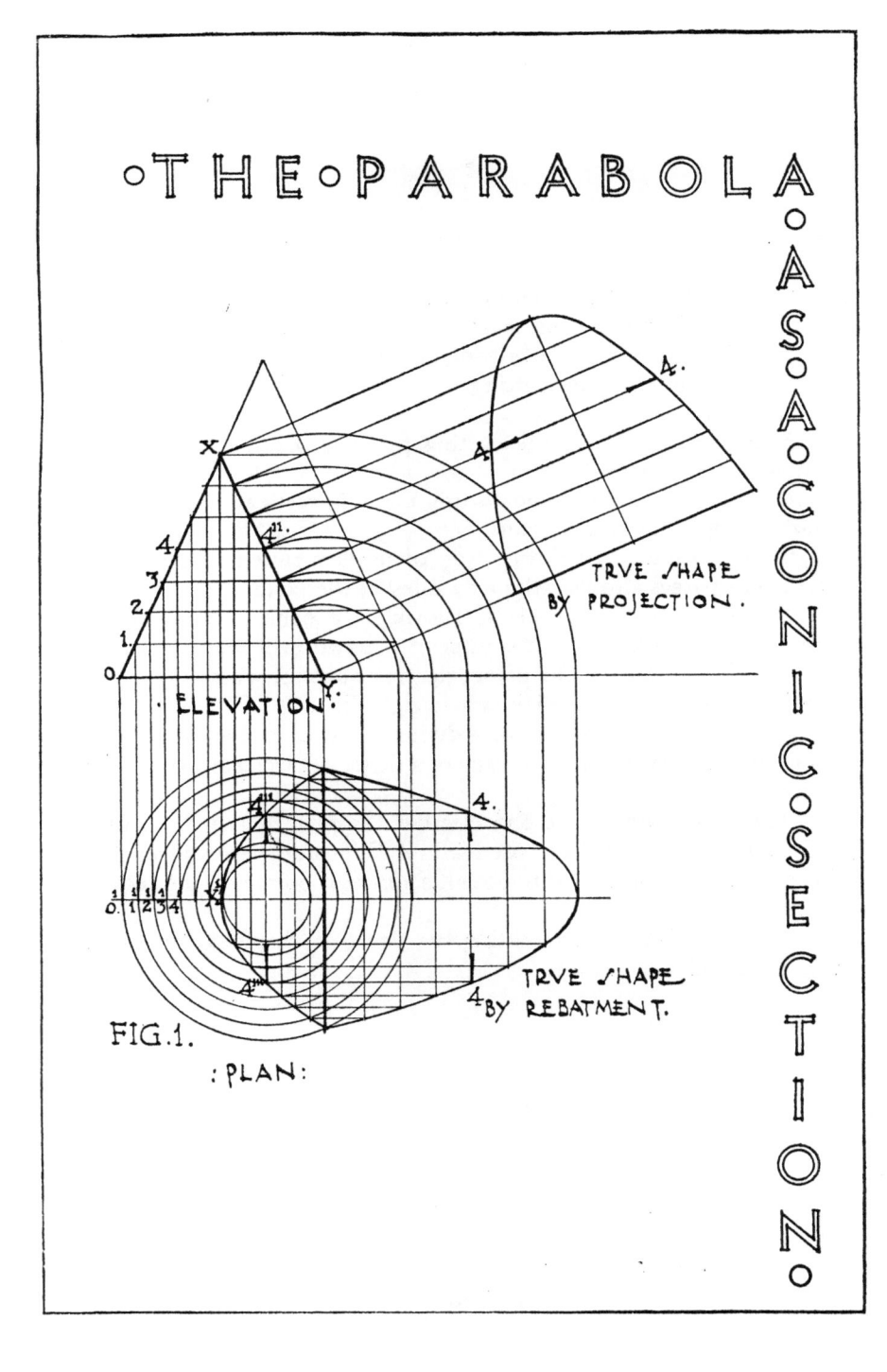

THE PARABOLA AS A CONIC SECTION

TRUE SHAPE BY PROJECTION.

ELEVATION.

TRUE SHAPE BY REBATMENT.

FIG. 1.

:PLAN:

THE HYPERBOLA AS A CONIC SECTION

A CONE may be cut by planes in various ways, as shown by Fig. 1, and the resulting sections have different outlines:

1. A *triangle* when the cutting is through the vertical axis.
2. A *circle* when the cutting is horizontal, at right angles to the vertical axis.
3. A *parabola* when the cutting plane is parallel to one of the sides of the cone.
4. A *hyperbola* when the cutting plane is vertical, parallel to but not through the vertical axis.
5. An *ellipse* when the cutting plane is inclined to the vertical axis and passes through the two sides of the cone.

In Fig. 2 the hyperbola is shown as a conic section and the method of construction is similar to the two previous examples.

Let X-Y be the cutting plane, the true shape of which is required. As X-Y is vertical the plan is a straight line.

By means of horizontal lines divide the vertical height X-Y into a number of parts and by means of a series of circles draw the plans of these lines.

The points where these circles cut the plan of the plane determine a series of widths in the section. Transfer these widths to the elevation lines produced to obtain a series of points on the required curve.

The true shape of the cutting has also been found by rebatment, using point X as centre.

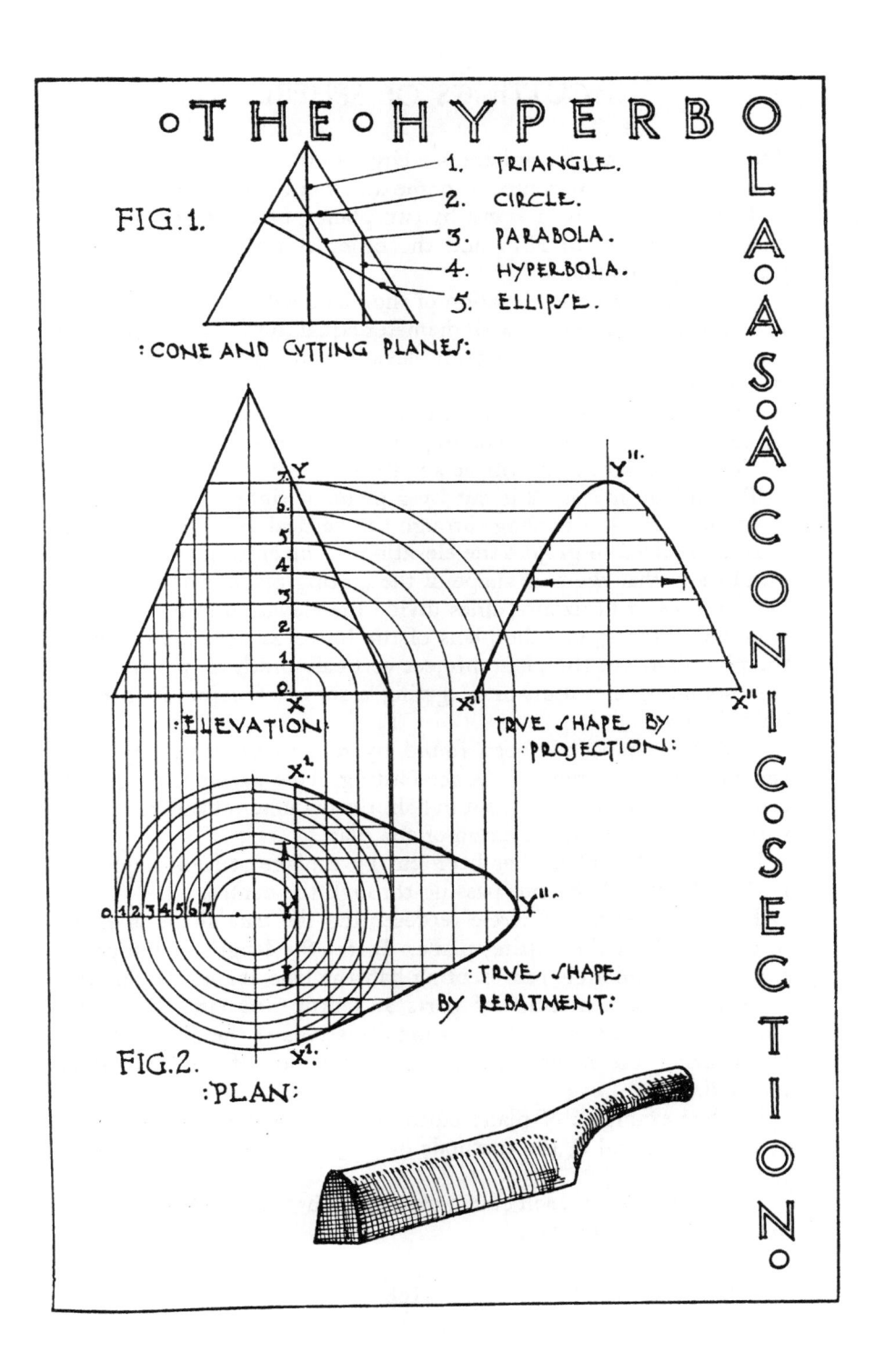

THE HYPERBOLA AS A CONIC SECTION

FIG. 1.

1. TRIANGLE.
2. CIRCLE.
3. PARABOLA.
4. HYPERBOLA.
5. ELLIPSE.

:CONE AND CUTTING PLANES:

:ELEVATION:

:TRUE SHAPE BY PROJECTION:

:TRUE SHAPE BY REBATMENT:

FIG. 2. :PLAN:

CUTTINGS OF SPHERE

ANY section through a sphere is a circle in true shape. This fact is of considerable importance in dome and vault construction.

The sphere in Fig. 1 is cut by two planes, one at right angles to V.P. and parallel to H.P. and the other at right angles to both planes of projection.

In the first case the elevation of the plane will show as a straight line and the plan a circle with diameter equal to the elevation of the plane. In the second case both elevation and plan will appear as straight lines.

The inclined view of a circle is an ellipse, so that any view of a plane cutting through a sphere, other than one that produces a circle or a straight line, will be an ellipse.

The sphere in Fig. 2 is cut by a plane at right angles to H.P., at 45° to V.P., and passing through the vertical axis of the sphere.

It is required to project the elevation of the cutting and to prove graphically that the true shape of the cutting plane is a circle.

By means of horizontal lines divide the elevation into a number of parts, drawing also the plans of the lines. At the points where the plan lines cut the plane project vertically until they meet their corresponding elevation lines, giving a series of points in the required curve.

The true shape has been found by rebatment using any conveniently situated centre. As the cutting plane passes through the centre of the sphere and cuts it in half, the diameter of the true shape will equal that of the elevation or the plan.

In Fig. 3 the sphere has been cut by a plane at right angles to H.P. and 60° at V.P., not passing through the centre.

The diameter of the circle representing the true shape is equal to the length of the cutting plane. Bisect the length of the plane and erect a semicircle representing half the true shape. Divide the semicircle into a number of parts and drop ordinates on to the diameter as at X and Y. Project the elevations of the division lines and transfer the heights X, Y, etc., either side of the horizontal centre line.

Further examples of plane cuttings through spheres are shown in Figs. 4, 5, and 6, the methods of projection being the same in each case.

The true shape of each cutting is a circle having a diameter equal to the length of the plan.

CUTTINGS OF SPHERE

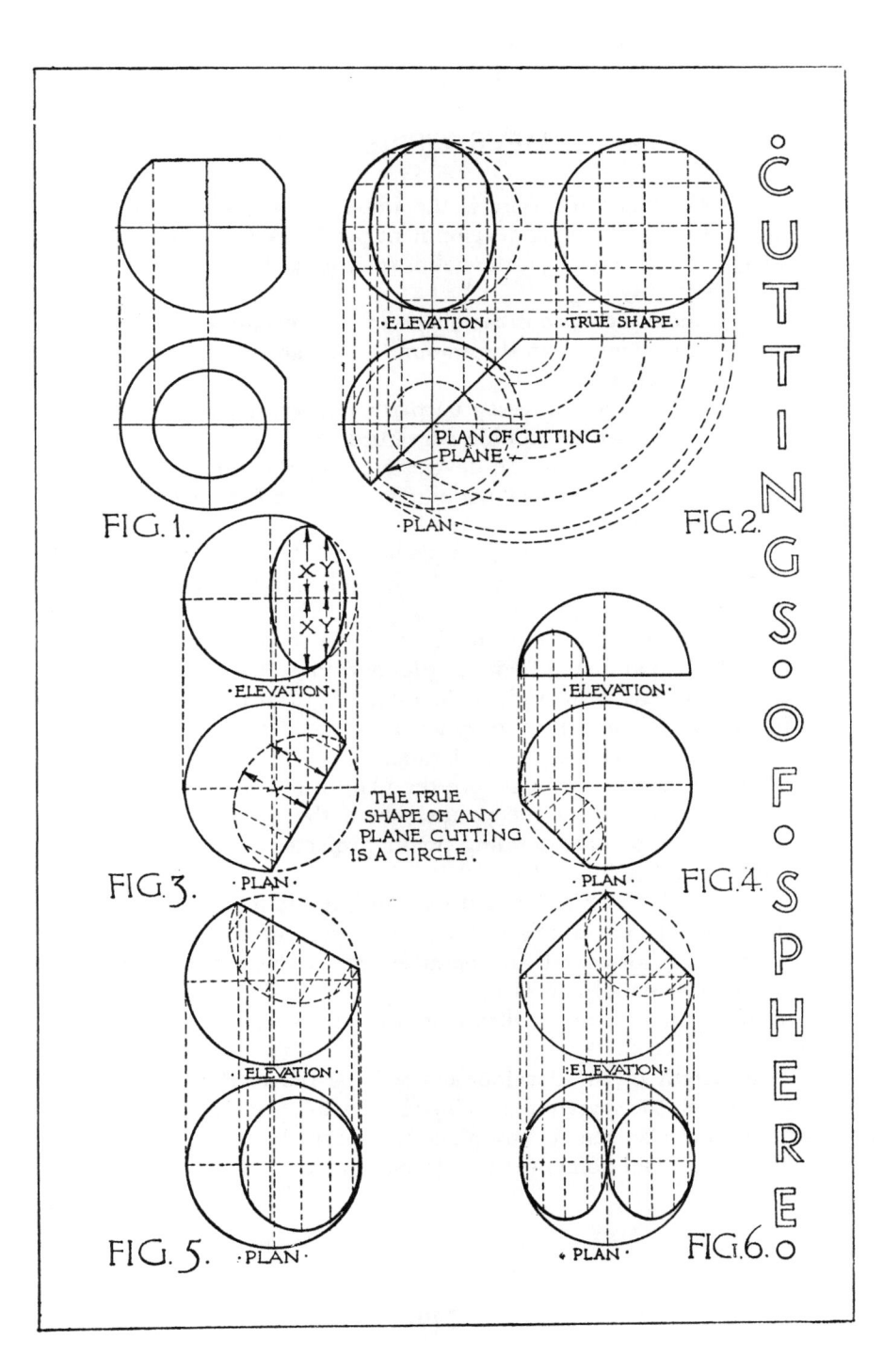

FIG. 1.

FIG. 2.
- ELEVATION
- TRUE SHAPE
- PLAN OF CUTTING PLANE
- PLAN

FIG. 3.
- ELEVATION
- PLAN
- THE TRUE SHAPE OF ANY PLANE CUTTING IS A CIRCLE.

FIG. 4.
- ELEVATION
- PLAN

FIG. 5.
- ELEVATION
- PLAN

FIG. 6.
- ELEVATION
- PLAN

DEVELOPMENTS

The development of a solid is the amount of its surface area projected into one plane. A development produces a mould or pattern that completely envelops the solid, or it may be a development of one surface only.

In practical setting-out the shape of the development is usually of more importance than the amount of the area, which can nearly always be calculated.

Developments are an essential part of sheet metal work, also of constructional work of double curvature.

Simple cases such as the development of the surfaces of cubes and prisms need little explanation; the cylinder has been chosen as our first example.

A cylinder, Fig. 1, is cut by a plane at 30° to H.P. It is required to develop the curved surface of the cylinder below the cutting plane.

Divide the circumference of the cylinder in plan into a number of equal parts and step their lengths out on a development line as at 1_1, 2_1, 3_1, etc., which gives the total length of the development. This is not, of course, quite exact, but providing sufficient points are taken the inaccuracy is negligible.

Draw vertical projectors from the plan to the elevation and project horizontally from the points where the projectors meet the cutting plane to corresponding points projected from the development as at $4''-4'''$ and $10''-10'''$.

The cylinder in Fig. 2 is cut by two intersecting planes.

A similar procedure is followed and part of the development appears as a horizontal line because one of the cutting planes is horizontal or parallel to H.P.

A hexagonal prism is shown in Fig. 3, cut by a plane at 45° to H.P.

The length of the development will be the total length of the six sides of the pyramid as shown in plan. As these sides are straight and also the cutting plane is straight there is no object in taking additional points for the development.

:ELEVATION: :DEVELOPMENT:

:PLAN:

FIG.1.

:DEVELOPMENT OF SURFACE OF CYLINDER:

CUTTING PLANES

:PLAN:

FIG.2.

CUTTING PLANE.

:ELEVATION: :DEVELOPMENT OF SURFACE:

FIG.3.

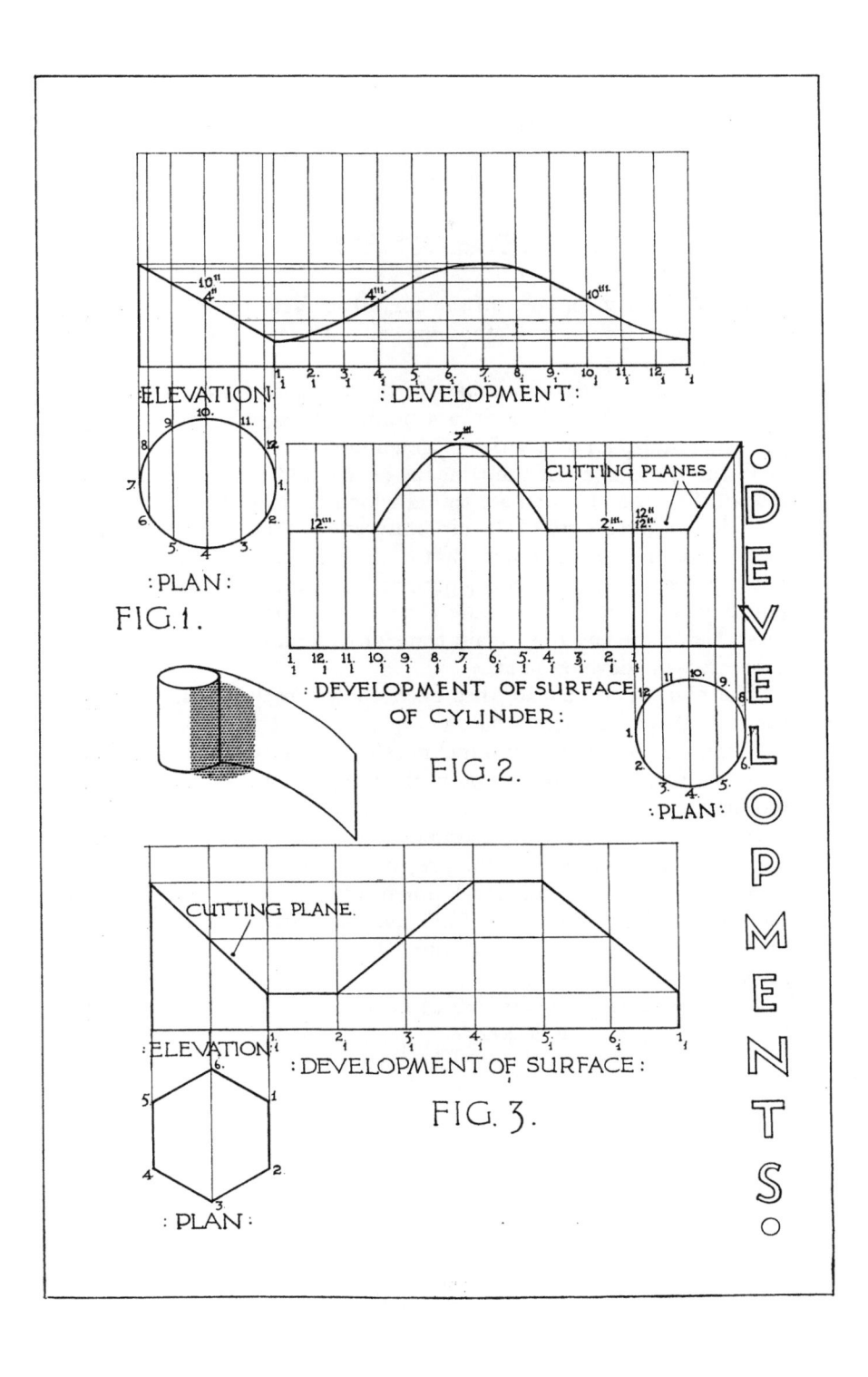

:PLAN:

DEVELOPMENTS

DEVELOPMENTS

In Fig. 1 the development of the curved surface of a cone is shown, also the development of the line made by an inclined cutting through the cone.

Draw the elevation and plan of the cone, and by dividing the circumference of the base into a number of equal parts draw the elevation and plan of a series of generators.

With O, the apex of the cone, as centre and radius equal to the side of the cone, draw an arc round which are stepped the divisions 1, 2, 3, etc., the length of the circumference of the base of the cone.

To find the development of the line the cutting plane makes with the surface of the cone, project horizontally to side O–1 the points where the cutting plane meets the elevation of the generators. Using the centre O again transfer these distances to the same generators in development.

The plan of the cutting has been shown in the illustration but the true shape has been omitted.

A complete exercise could include this, and it would be necessary to do so if the development of all the surfaces of the cut cone were required.

In Fig. 2 an octagonal pyramid has been treated in a similar manner as was the cone in the previous example.

It will be seen that the development of each of the eight sides of the pyramid is a triangle. The division lines 1, 2, 3, etc., are measured on the base stepped round the arc and joined by straight lines.

As the cutting plane cuts eight faces the development line will consist of eight straight lines.

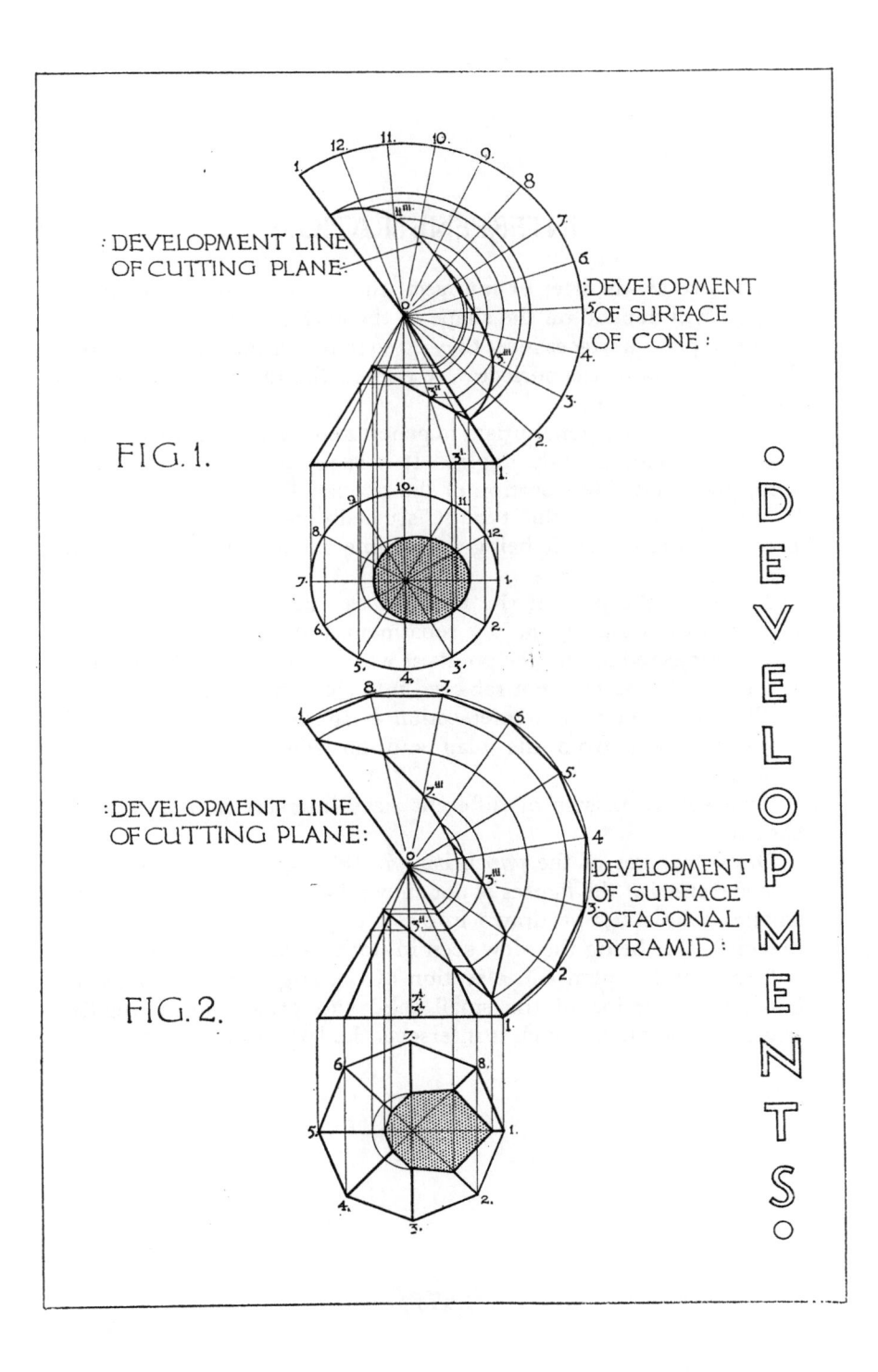

FIG. 1.

DEVELOPMENT LINE OF CUTTING PLANE.

DEVELOPMENT OF SURFACE OF CONE.

FIG. 2.

DEVELOPMENT LINE OF CUTTING PLANE:

DEVELOPMENT OF SURFACE OCTAGONAL PYRAMID:

DEVELOPMENTS

INTERPENETRATION

WHEN two solids meet or intersect, the shape of the line of intersection will depend on the shape of the surfaces in contact.

If two plane surfaces meet they do so in one or more straight lines. If either or both the surfaces are curved the intersection line will be a curve or curves.

In Fig. 1 a hexagonal prism is penetrated by an octagonal prism, their axes being at right angles. It is required to draw the elevation, plan, and side elevation of the two solids.

With problems of this type it is usually necessary to draw two projections together, it being impossible to complete one without the other.

In Fig. 1 the plan of the hexagon can be drawn and the side elevation of the octagon. By rebatment the plan of the octagon can be projected to find the points at which it intersects the hexagon and similarly the hexagon rebatted into side elevation.

The elevation of the intersection is found by projecting each point vertically from the plan and horizontally from the side elevation.

Two square prisms of different sizes intersect in the example shown in Fig. 2.

Draw the axes of the two prisms in their given positions. The side elevation of the figures is not shown but the true section of the smaller solid must be drawn in order to obtain the length of the diagonal, this being the view seen in both elevation and plan.

Complete the plan and elevation of the large prism, and by rebatting the section of the small solid and projecting its angular edges the points at which it intersects the large solid are obtained.

FIG 1 ·ELEVATION· ·SIDE ELEVATION· :PLAN

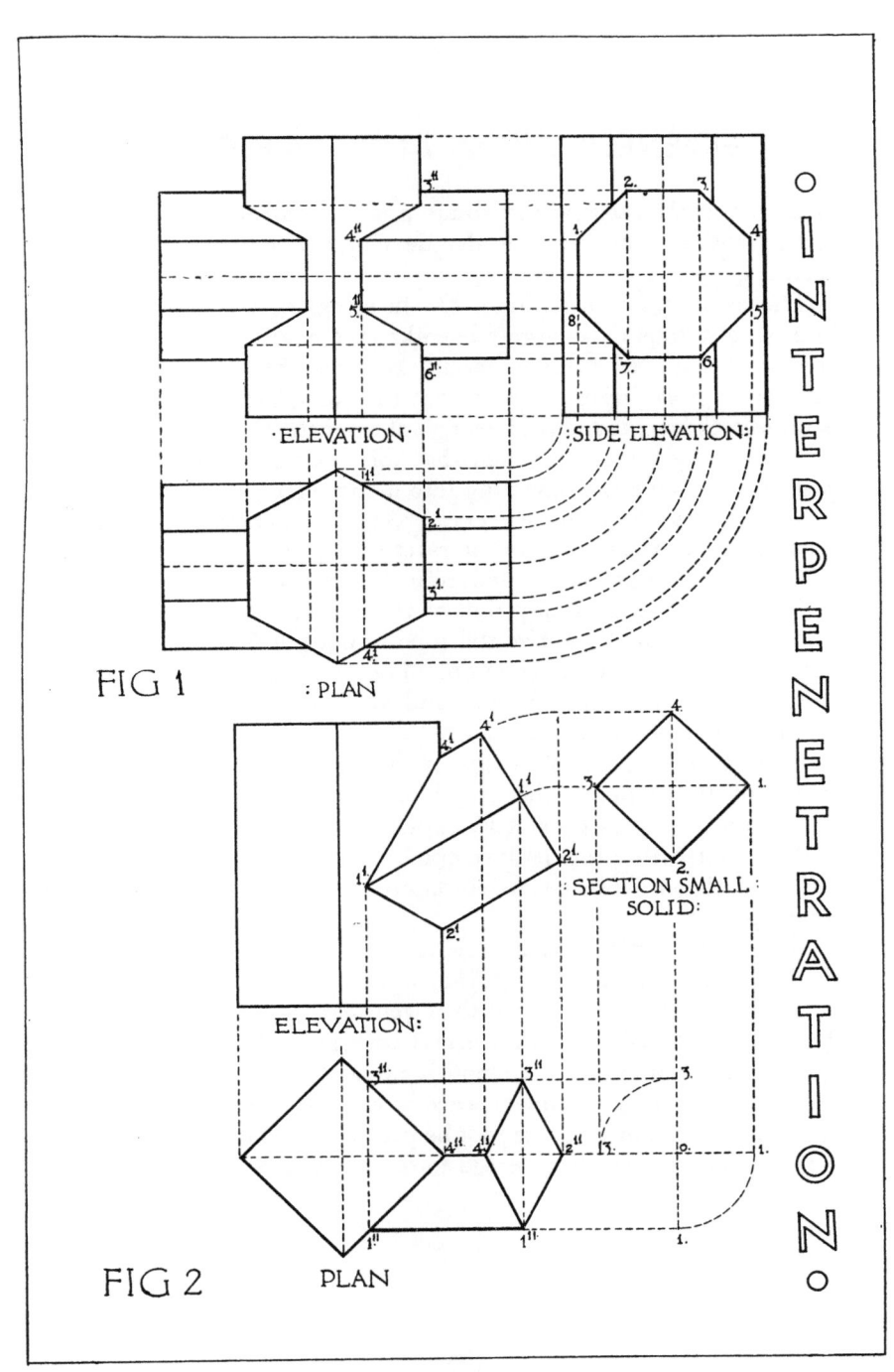

FIG 2 ELEVATION: ·SECTION SMALL SOLID· PLAN

INTERPENETRATION

INTERPENETRATIONS AND DEVELOPMENTS

It is frequently necessary to obtain not only various views of intersecting solids or figures, but also developments of the intersecting surfaces.

The examples shown under the headings of developments and interpenetrations have largely been kept separate, but have been so arranged that the student should be able, by reference to the development examples, to work out the constructions necessary to complete those examples showing interpenetrations only.

It is strongly recommended that this should be done wherever possible, to make each drawing into a complete exercise.

The two examples shown here are of interpenetrations and developments as they occur in intersecting pipes.

In Fig. 1 two pipes of equal radii intersect at right angles.

Draw the plan and elevation of the pipes, and by dividing the circumference of the horizontal pipe into a number of equal parts project the lines of intersection. The two pipes being of equal radii, on the same vertical axis and at right angles, these will be straight lines in elevation.

Treating each pipe as a cylinder the development of the surface is found by stepping the length of the circumference along a development line as previously explained.

The intersection line is developed by projecting the intersection of the division ordinates in elevation onto the same ordinates in development.

In Fig. 2 the pipes are of unequal diameter and are not at right angles to each other although they are on the same vertical axis.

To obtain the development the circumferences of the pipes must be divided separately as they are of unequal radii.

The development of the small pipe has been placed clear of the rest of the drawing for convenience of spacing, therefore the heights on each division ordinate cannot be projected but must be measured on the elevation and transferred to the development as at X and Y.

HALF DEVELOP
OF HORIZ.
PIPE.

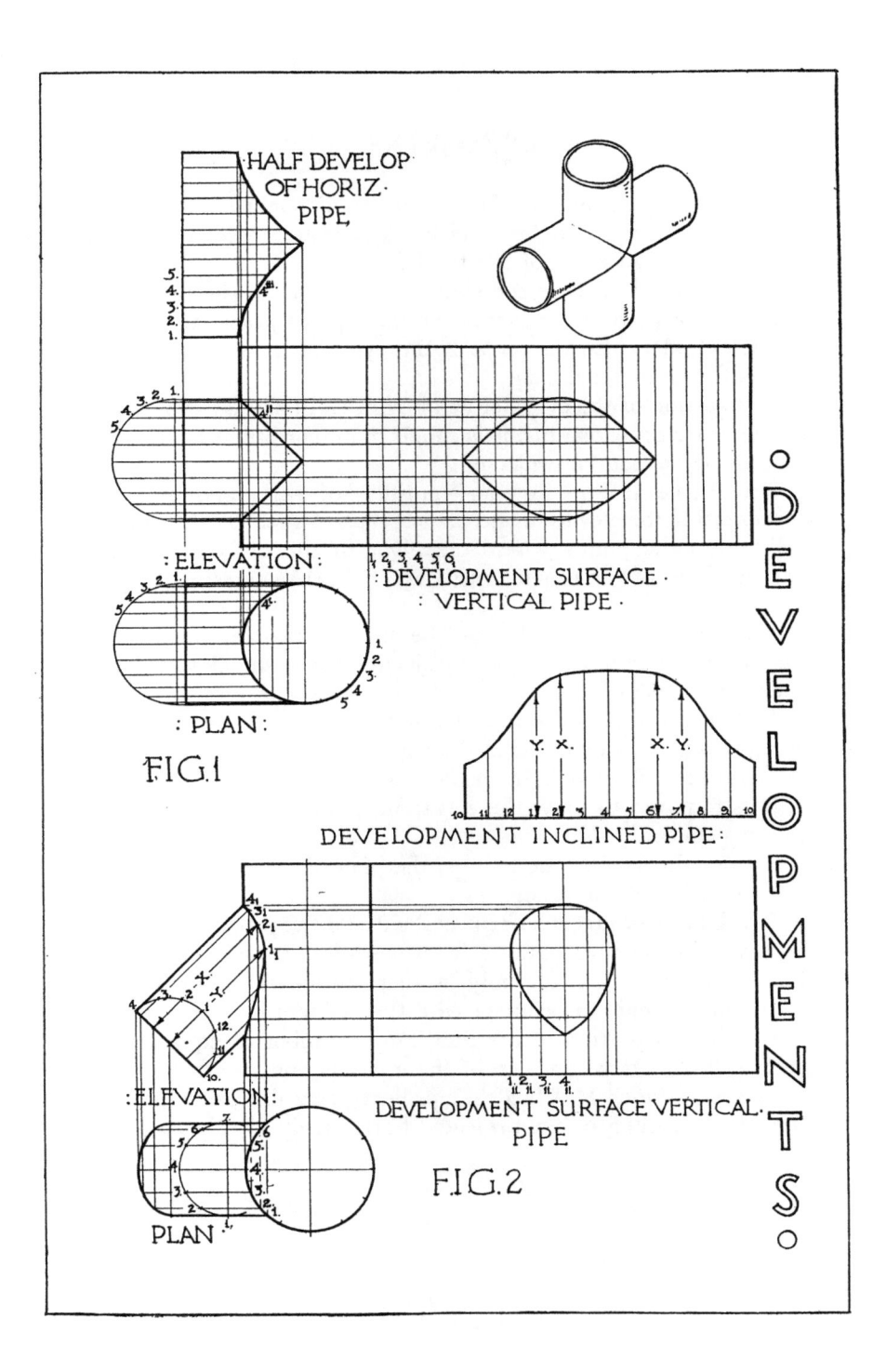

:ELEVATION: :DEVELOPMENT SURFACE:
              : VERTICAL PIPE :

: PLAN :

FIG.1

DEVELOPMENT INCLINED PIPE:

:ELEVATION: DEVELOPMENT SURFACE VERTICAL
              PIPE

PLAN   FIG.2

DEVELOPMENTS

INTERPENETRATION

A SQUARE prism stands on H.P. with its diagonal parallel to V.P. and its axis at right angles to H.P. A cylinder passes through the prism making an angle of 30° to H.P., Fig. 1.

It is required to draw the elevation and plan of the two solids.

Draw the plan and elevation of the square prism and the axis of the cylinder, also the ends of the cylinder in elevation at right angles to the inclined axis.

The section of the cylinder can be drawn in any position relative to the other drawings but it is much more convenient if placed so that all points can be projected.

Divide the section into any number of parts and draw lines representing the elevation and plan of these divisions. Project vertically from the points where the plans of the divisions meet the vertical faces of the prism until they meet the corresponding lines in elevation.

The ends of the cylinder will be ellipses in plan, and points in the curve are found by direct projection from the elevation using the same division lines.

In Fig. 2 a cylinder and a hexagonal prism intersect each other at right angles. The prism is the larger of the two solids, therefore the cylinder will pass right through it.

It is required to draw the elevation, plan, and side elevation of the two solids.

Set out the axes of the figures, the plan of the cylinder, and the side elevation of the prism.

It will be seen that the horizontal sides of the prism in side elevation will remain horizontal in front elevation, therefore the curved intersection will be made only by that portion of the inclined side of the prism that is in contact with the cylinder.

Divide this portion of the line into a number of equal parts as at 1, 2, 3, 4. Draw the plan of these points and where each intersects the cylinder 1', 2', 3', 4', project vertically until they meet the corresponding points projected horizontally 1", 2", 3", 4".

INTERPENETRATION

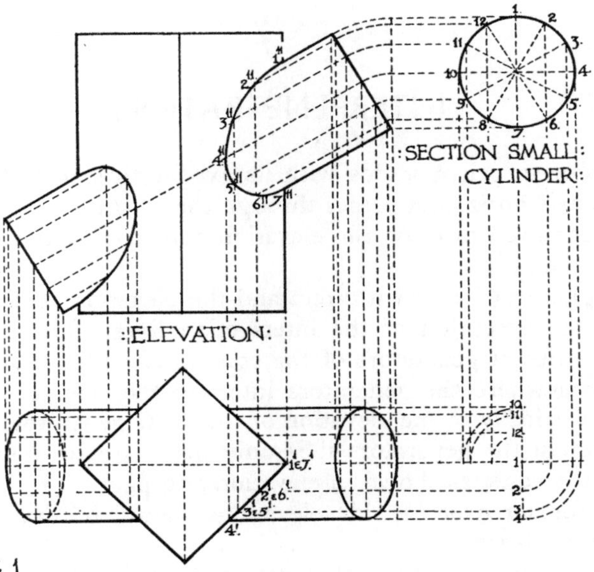

FIG 1

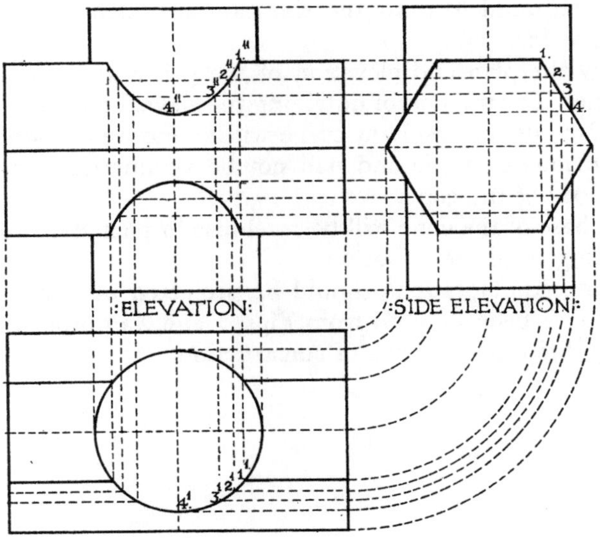

FIG.2

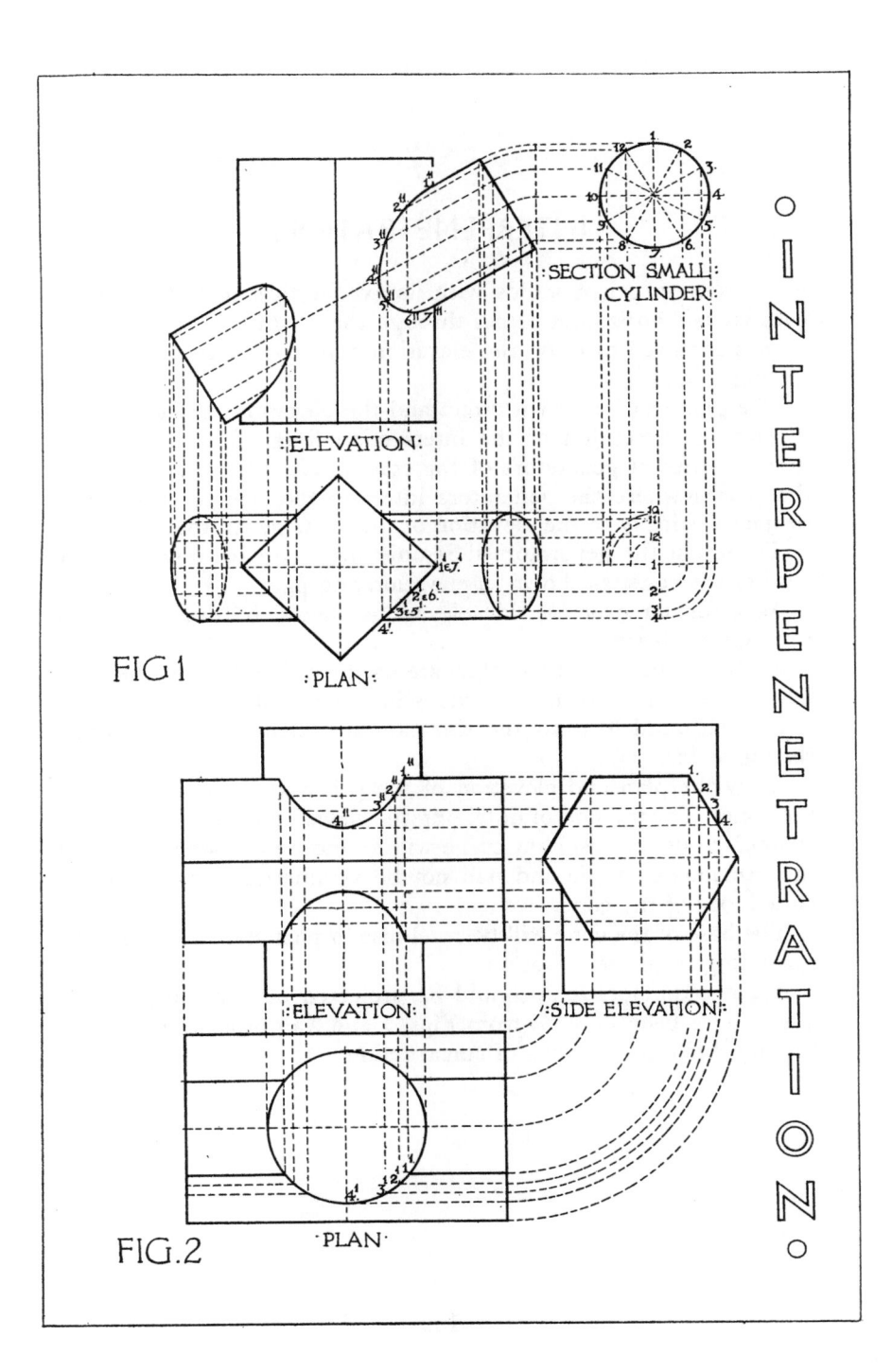

INTERPENETRATION

A HEXAGONAL prism stands with its axis vertical to H.P. A cone with its axis horizontal passes through the prism.

It is required to draw the elevation and plan of the two intersecting solids.

The plan can be drawn direct and the elevation projected from it with the exception of the intersection lines. To obtain these draw a series of generators of the cone in elevation and plan. At the points where the generators intersect the sides of the prism project vertically to the elevation of the same generators.

To obtain the generators it is only necessary to draw half the base of the cone, as shown in elevation and plan.

As a further exercise side elevations should be drawn looking from either direction.

In Fig. 2 the same two solids are used; the position of the cone has been changed so that its axis is inclined at 30° to H.P.

It is required to draw the elevation and plan of the two intersecting solids.

Draw the plan and elevation of the prism and the elevation of the cone. The method of obtaining the elevation of the intersection is similar but the resultant curve will be more pronounced than in the previous example and will not be symmetrical about the inclined axis of the cone.

The base of the cone will be an ellipse in plan and is obtained by projecting from the elevation.

Again, side elevations should be projected as drawing practice. These have been omitted from Figs. 1 and 2 in order to introduce slightly different methods of construction.

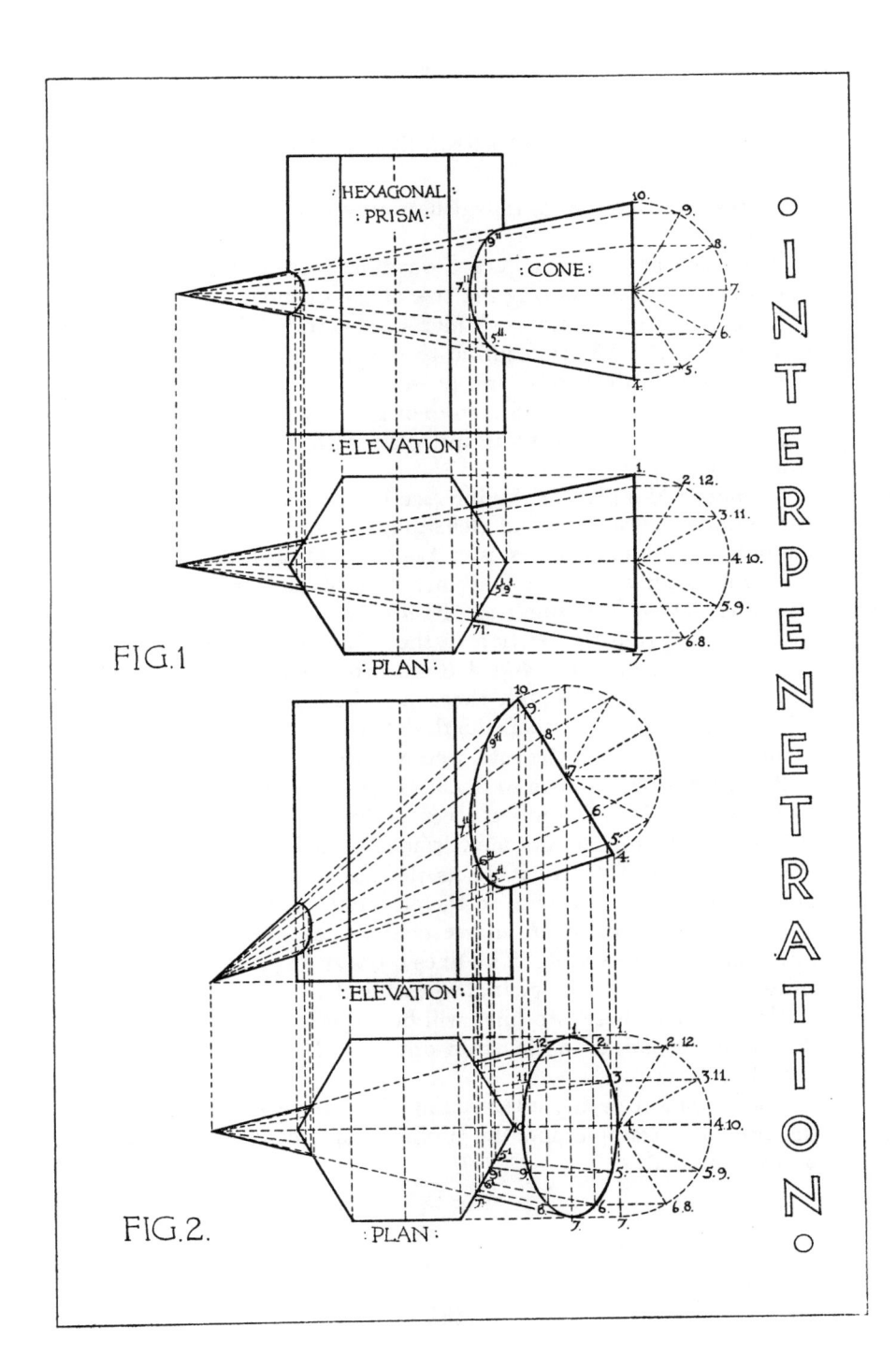

INTERPENETRATION

FIG.1 — HEXAGONAL PRISM : CONE : ELEVATION : PLAN

FIG.2 — ELEVATION : PLAN

INTERPENETRATION

Two further examples of interpenetrating solids are shown in Figs. 1 and 2.

The cone and cylinder are the solids chosen and the exercise has been varied by altering the size of the cylinder.

In Fig. 1 a horizontal cylinder is penetrated by a right cone standing on H.P. It is required to find the line of intersection of the two solids in elevation and plan.

Draw the axes of the solids in their given positions and complete the elevation, end elevation, and plan with the exception of the intersection lines.

To obtain the plan of the intersection line divide the section of the cylinder, as seen in end elevation, into a number of parts and connect by horizontal lines as at 12–2, 11–3, 10–4, etc.

It will readily be seen that that part of the circle between X and Y is clear of the cone and therefore makes no line of intersection. Considering the portion of the cylinder that does touch the cone, if horizontals 9–5 and 8–6 are produced and considered as horizontal cuttings through the cone, their plans will be circles. This point where the circle cuts the plan of the division line on the cylinder is a point on the intersection curve.

Point 5 has been shown in detail. Project the point to the side of the cone, 5', drop it onto the centre line in plan, 5″, and draw the plan circle until it cuts the horizontal line in plan, 5‴. Project the point vertically until it meets the horizontal projection of the point giving 5″″ as a point in the elevation of the curve.

In Fig. 2 the cylinder intersects the cone giving straight-line intersections in elevation. The lines cross at P, the point where the sides of the cone are tangential to the circle.

The plans of these two lines will be ellipses and are obtained by dropping the points in elevation down to the plans of the corresponding lines.

The elevation of the intersection could have been found as in the previous example and the problem should also be attempted by that method.

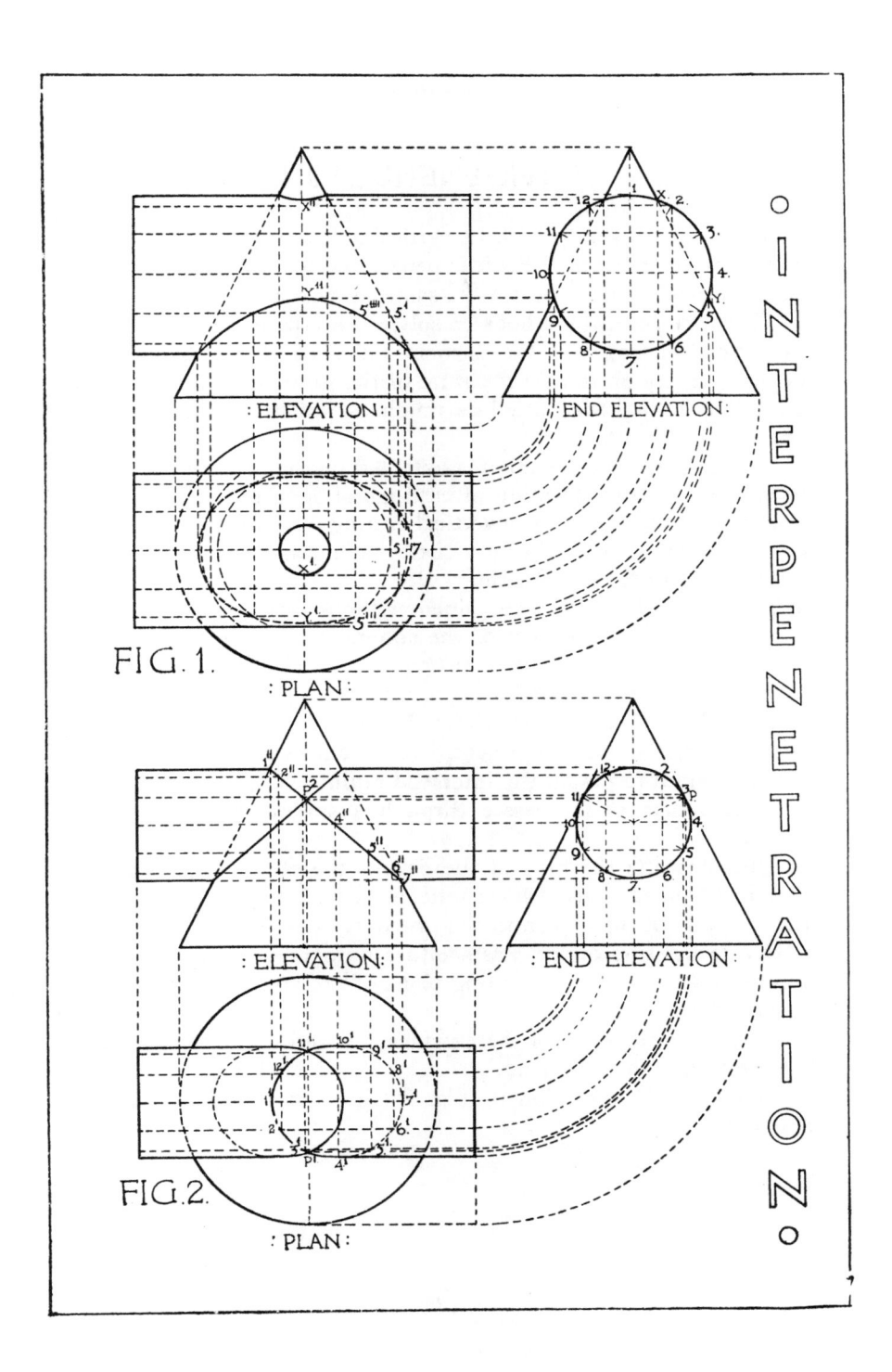

: ELEVATION : : END ELEVATION :

FIG. 1. : PLAN :

: ELEVATION : : END ELEVATION :

FIG. 2. : PLAN :

INTERPENETRATION.

INTERPENETRATION

A SQUARE prism stands with its vertical axis at right angles to H.P. and its sides at 45° to V.P. It is penetrated by a sphere having the same horizontal axis but not the same vertical axis.

There are several methods of solving interpenetration problems involving spheres, and the particular line of approach used will depend largely on the exact nature of the problem. Here the constructions have been varied as much as possible to demonstrate these different methods.

In Fig. 1 the side elevation is obtained by the method previously used to obtain the elevation of cuttings of spheres. If the side of the prism is considered as a vertical cutting plane the true elevation of the intersection would be a circle having as diameter that part of the line in contact with the circle.

The front elevation is obtained without reference to the side elevation. Divide the plan of the sphere into a number of radial divisions and treat each as a cutting plane. The elevation in each case will be a semi-ellipse having as major axis the vertical diameter of the sphere. The points where the prism in plan intersects the cutting planes should be projected to the elliptical elevations of the cutting planes to obtain the intersection line.

A triangular prism passes through a sphere in the example, Fig. 2.

After drawing the plan of the figures the other projections are obtained from a series of horizontal cutting planes, as distinct from the vertical cuttings taken in the previous example.

Divide the side of the prism into a number of parts and using centre O draw arcs representing plans of horizontal cutting planes passing through these points.

Project the elevation of these planes and onto them the points of intersection with the prism.

Note that the elevation in Fig. 2 has been completed on one side only.

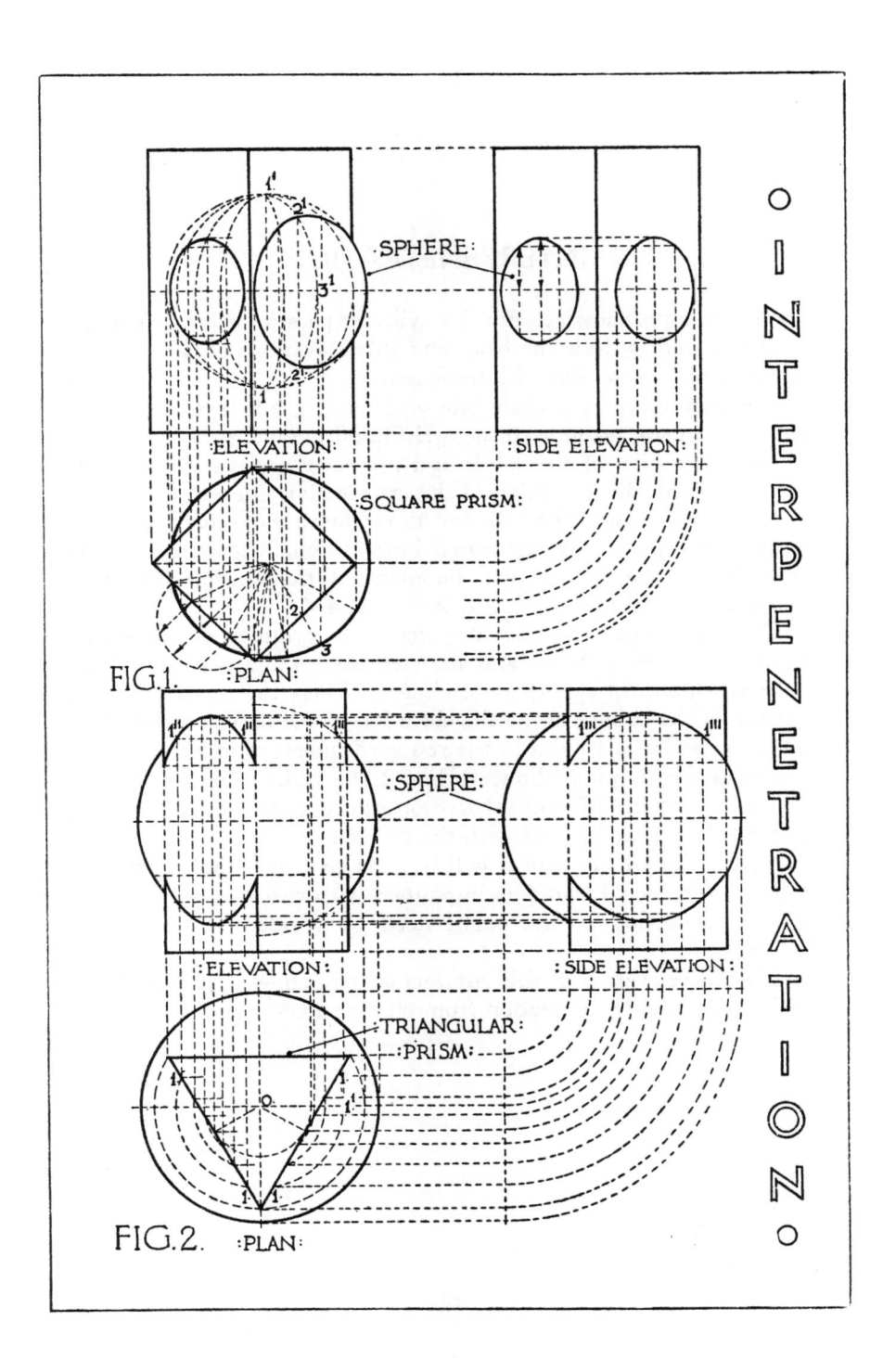

FIG.1.

ELEVATION — SPHERE — SIDE ELEVATION

PLAN — SQUARE PRISM

FIG.2.

ELEVATION — SPHERE — SIDE ELEVATION

PLAN — TRIANGULAR PRISM

INTERPENETRATION

INTERPENETRATION

In the example shown in Fig. 1 a cylinder passes through a sphere. It is required to draw the front and side elevations of the solids in order to obtain the line of intersection.

The method used is again one of "slicing", but in this example the cuttings are taken vertical to H.P. and parallel to V.P.

Draw the plan of the solids and project the elevation and side elevation with the exception of the intersection line.

Divide the plan of the cylinder into a number of equal parts and connect the points by horizontal lines which, when produced to the plan of the sphere, will be plans of the cutting planes or "slicings" parallel to V.P., 1-1, 2-2, 3-3, 4-4.

Draw the elevation of the divisions on the cylinder, which will be a series of vertical lines, and the elevation of the cuttings of the sphere, which will be a series of circles. Where the elevation of the cutting of the sphere meets the elevation of the corresponding line on the cylinder is a point in the required intersection curve.

A similar example is shown in Fig. 2, the variation being that the two solids are in different relative positions, so that only a portion of the sphere is in contact with the cylinder.

Having drawn the plan it will be seen that only that portion of cylinder between 1 and 6 is in contact with the sphere, therefore only that portion of the curve need be divided to obtain the "slices".

The side elevation of the intersection will show as a dotted line and is obtained by projection from elevation and plan.

INTERPENETRATION

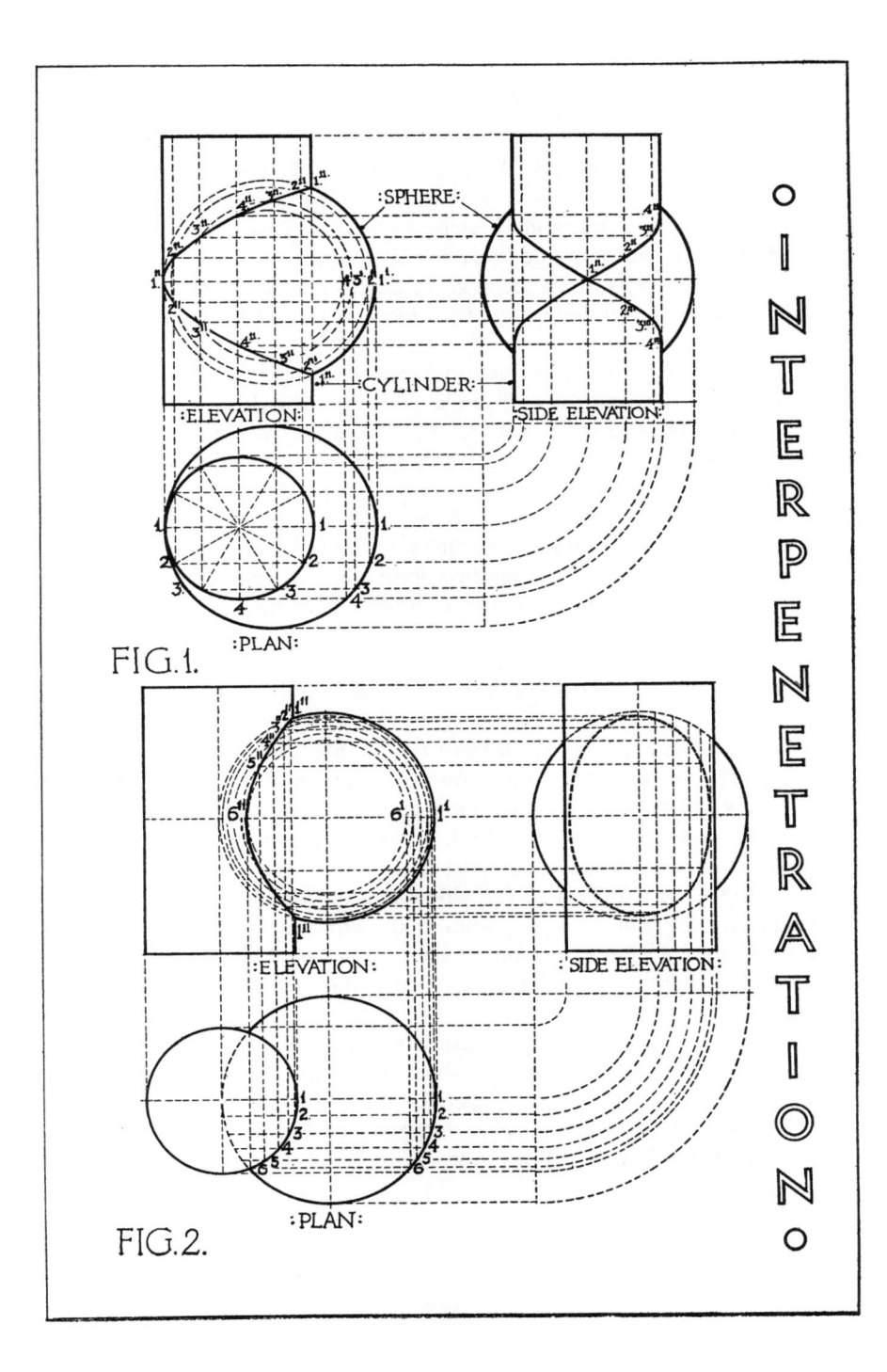

FIG. 1.

:SPHERE:
:CYLINDER:
:ELEVATION:
:SIDE ELEVATION:
:PLAN:

FIG. 2.

:ELEVATION:
:SIDE ELEVATION:
:PLAN:

INTERSECTING MOULDINGS

Two practical examples of intersecting solids are shown in Figs. 1 and 2.

In Fig. 1 the horizontal member is the transom of a mullion and transom window and the vertical member a mullion.

The mouldings intersect at right angles and are similar sections; therefore at the underside of the transom the intersection line will be a 45° mitre.

The top edge of the transom is weathered to throw off the water that would fall upon it and is also rebatted for the window frame. It is the intersection of the vertical mouldings of the mullion with the weathered surface of the transom that creates the geometrical problem.

Draw the plan of the mullion and section of the transom, making the two moulded sections agree. By rebatting the end elevation of the mullion, together with some intermediate points in the moulding curves, obtain the points where the mouldings meet the weathering.

Horizontal projectors from these points meeting vertical projectors from the plans of the points give the intersection line in elevation.

The example in Fig. 2 is the type of intersection that occurs at the base of a moulded jamb stone stopping on a weathered stop.

Draw the plan of the jamb moulding and a normal section of the stop at its greatest depth, which in this case is through the centre of the casement member, on the line X—X.

Using C as centre find the side elevation of the moulding lines including some additional points in the curves. Project into elevation from the plan also from the points where the moulding lines meet the weathered surface to obtain the elevation of the intersection.

The small moulded base in front of the jamb does not intersect the mouldings but returns parallel to the face and to the reveal, forming a splay stop in each case.

The elevation of the stop is obtained by direct projection from plan and section but does not show as the true shape of the moulding.

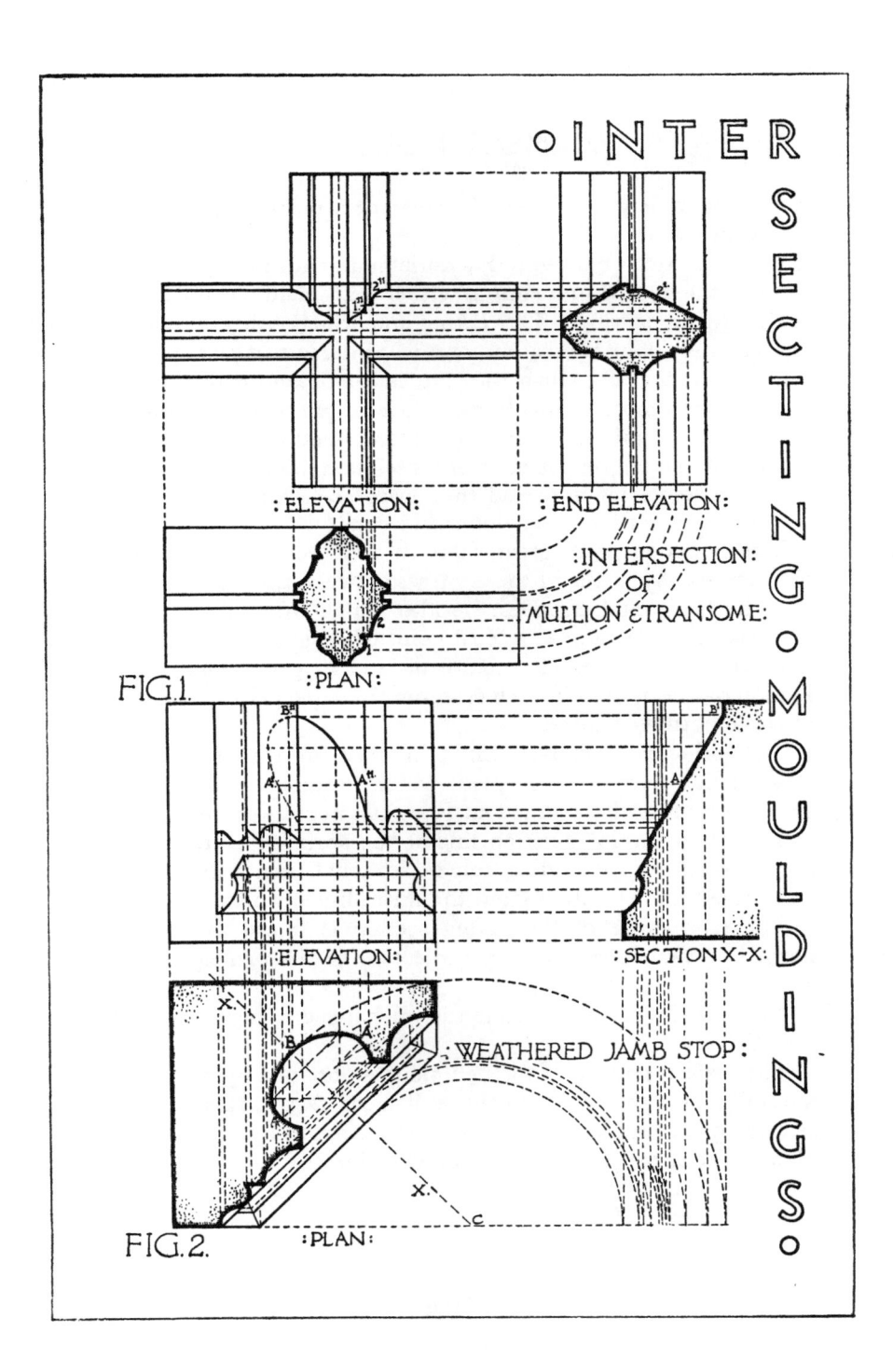

INTERSECTING MOULDINGS

FIG. 1. Intersection of Mullion & Transome — Elevation, End Elevation, Plan

FIG. 2. Weathered Jamb Stop — Elevation, Section X-X, Plan

PEDIMENT MOULDINGS

A PEDIMENT may be either triangular or a portion of one or more circles in elevation.

That portion between the horizontal and raking cornices is termed the tympanum and is either left plain or finished with sculpture or carving. The raking mouldings whether curved or straight may continue as one cornice, intersect at the apex, or be left open at the top, when they are called broken pediments.

Pediments most probably owe their origin to the inclined roofs of primitive buildings. In classic architecture they are mainly triangular as seen on the ends of Greek temples over the porticoes.

In the Renaissance period they were used with varying broken or unbroken outlines over doors and windows.

The pediment is of great value in architectural composition to give variety, by virtue of the contrast between the raking or curved and the horizontal cornices; besides which the definite apex can emphasize an opening as a focal point of design.

Both the examples shown are of broken pediments; the same section has been used in each case and the same plan has been made to serve both elevations.

The principal members must intersect at the springing in a straight line mitre, 45° on plan, which produces a section that is wider normal to the inclined surface than normal to the horizontal.

To obtain the intersection line at the springing, draw the plan of the mouldings including some additional points in the curves, and project the elevation from the mitre line on plan.

The true shape of the raking section is found by projecting, normal to the slope, the plan projection of the principal member as at points 1, 2, 3, and 4.

The remainder of the section is unchanged being equal in both directions.

The broken portion of the pediment is obtained by returning the moulding at right angles to the wall face. Again, this will appear as a 45° mitre on plan and the elevation is found by projecting from plan to the corresponding lines in elevation.

PEDIMENT MOULDINGS

FIG. 1.

:BROKEN PEDIMENT:

:ELEVATION:

FIG. 2.

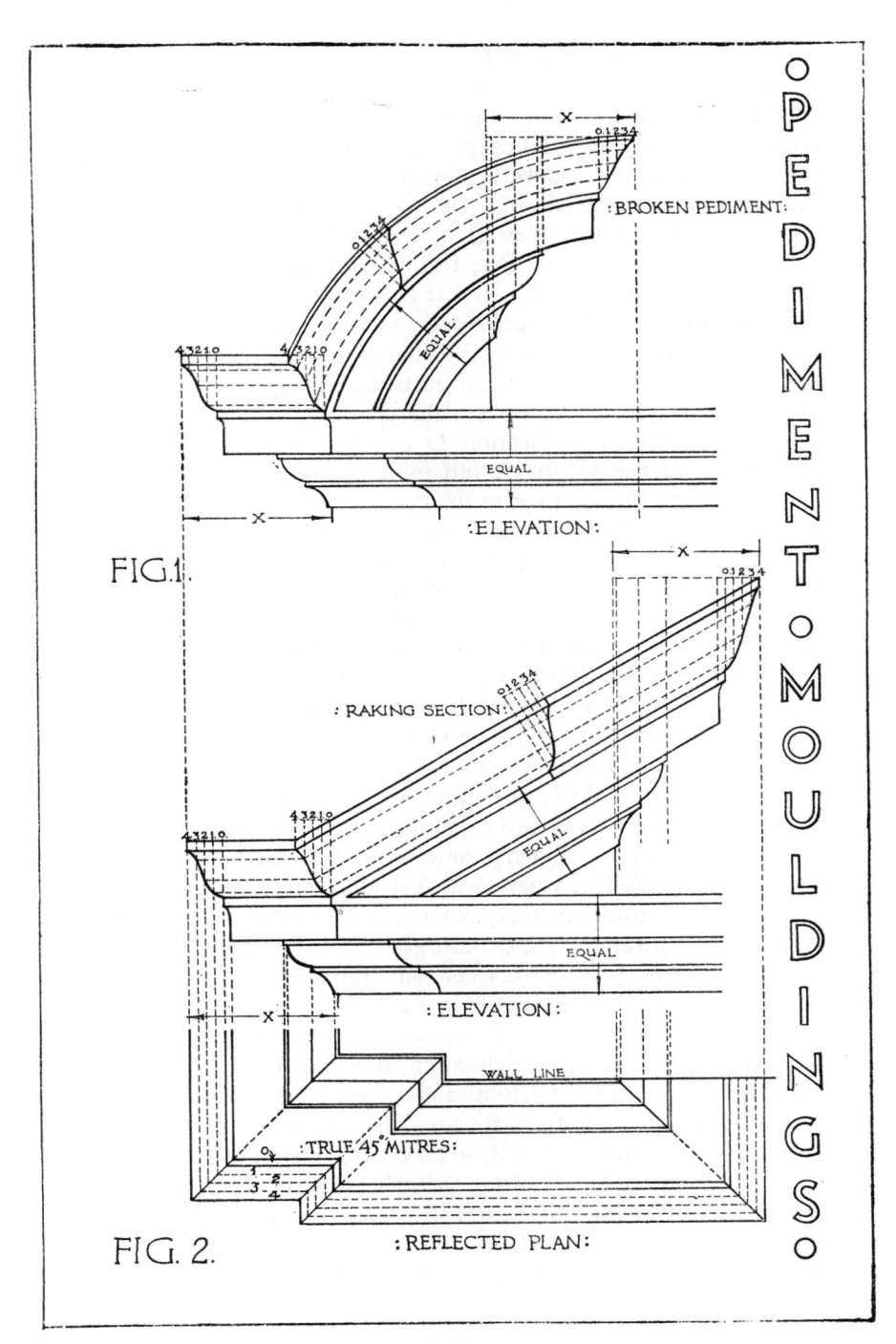

:RAKING SECTION:

:ELEVATION:

WALL LINE

:TRUE 45° MITRES:

:REFLECTED PLAN:

RAKING SECTIONS

PEMDIENTS are not the only constructional examples involving raking sections. Any continuous moulding which changes its inclination or its position relative to the vertical plane changes its section. Frequently mouldings change their direction relative to both the horizontal and vertical planes, as in staircase work when several different sections may be necessary in order to produce a continuous section intersecting in straight line mitres at the angles.

The example shown in Fig. 1 is of a moulding horizontal at A, B, D and E on plan and inclined at an angle of 30° to the horizontal at C. Sections A and B are the same and it is assumed here that B is the given section. Section C is obtained by precisely the same method as the raking pediment section, bearing in mind that the projection of the moulding remains constant throughout, but the heights are increased. Section D can be likened to the broken pediment and the section found in the same manner, projecting from the mitre line on plan to the corresponding lines in elevation. Both D and E being horizontal, the sections will be the same.

In Fig. 2 the moulding passes round a pier on plan whilst maintaining in elevation an angle of 30° to the H.P. Referring to the plan mouldings, A and D will be at 30° to the horizontal. Moulding B will be horizontal but the section will not be similar to A. Moulding C is inclined at 30° in elevation, but this is not its true angle of inclination to the H.P. To obtain the true shape of each moulding, assuming A to be the given section, complete the plan, keeping the projection P equal and indicating also some additional construction lines in the profile of the moulding. Find the horizontal section B by projecting from plan and elevation. Note that the angle of the top surface of the section has now become pitched at an angle of 120° to the front edge, whereas in A it was square.

Section D is similar in every respect to A but the section has been found again by assuming a cut at right angles to the slope in elevation, finding the plan of this cut, and rebating its true length in elevation into plan to find its true shape.

To find the true shape of section C, an auxiliary elevation is necessary. Find the true inclination of the moulding by rebating the length in plan to the height in elevation. Transfer this angle to the new ground line to form the auxiliary elevation. The moulding has to intersect with section B, so its vertical height at the mitre will be as h in section B. Transfer these heights to the new elevation and project from the mitre lines in plan to complete the elevation of the moulding. The true shape of the section is obtained by transferring the constant projection distance P to the new elevation lines obtained from h.

RAKING SECTIONS

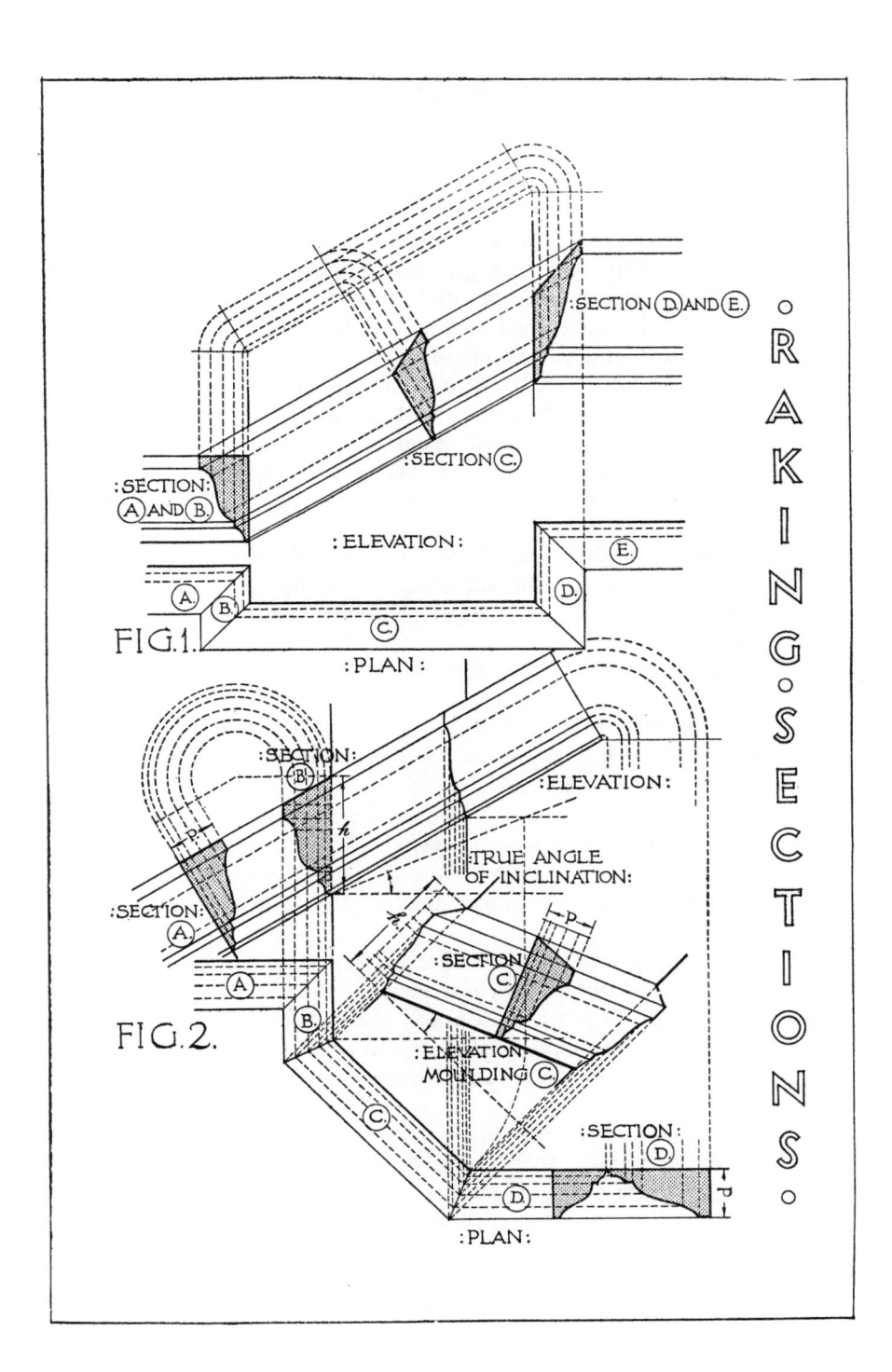

FIG. 1.

SECTION (D) AND (E).
SECTION (C).
SECTION (A) AND (B).
:ELEVATION:
(A) (B) (C) (D) (E)
:PLAN:

FIG. 2.

:SECTION: (B)
:ELEVATION:
:SECTION: (A)
(A)
(B)
TRUE ANGLE OF INCLINATION
:SECTION: (C)
:ELEVATION MOULDING (C):
(C)
(D)
:SECTION: (D)
:PLAN:

TRACERY

TRACERY originated in this country as a form of ornament, principally in windows, in the Norman period, but remained very simple until the introduction of the pointed arch about A.D. 1170. Earlier examples of Norman arcading with interlacing semicircular arches suggest tracery, in the same way that some consider this form of construction suggested the pointed arch.

The generally recognized origin was, however, "plate tracery", which consisted of a piercing through the tympanum formed when two arched openings occurred under one main arch.

The piercing was at first small and plain and usually in the form of a circle or lozenge. Foiled openings were introduced and the size of the opening increased until the width between the piercing and the arches of the openings equalled the width of the dividing mullion.

Bar tracery, as this construction is called, developed during the subsequent periods by the addition of varying forms of cuspings and of secondary systems of bars springing from the mouldings of the main bars.

The examples given here are intended to show the geometrical setting-out and the general principles of construction.

It will be seen that all the constructions have been previously studied under such headings as circles in contact, inscribed circles, tangents, and loci of centres.

The setting-out of tracery calls for particularly careful draughtsmanship to ensure the accurate separation of curves struck from different centres. Each curve must change direction at the point where it is cut by a common normal, the line joining the two centres concerned.

The system of setting-out the bars, the geometry of the problem, should be worked out on the centre lines of the mullions, adding the thickness of the bars next and finally the foils, cusps, and eyes.

In Fig. 1 the principle is that of the ogee arch constructed on equilateral triangles, the tracery being beneath a main equilateral arch.

In Fig. 2 there are two openings beneath a lancet arch. The centre for the circular opening has been found by locus of centres.

In Fig. 3 there are four openings beneath one equilateral arch. The same radius has been used throughout for the centre lines.

FIG. 1.

FIG. 2.   FIG. 3.

TRACERY

TRACERY PANELS

THE use of tracery as a form of ornamentation is by no means confined to windows, and during the Decorated and Perpendicular periods in particular it was extensively used for all forms of panelling. This type of tracery is often referred to as blank tracery as distinct from pierced tracery. The four examples illustrated could be used equally well either as blank panels or pierced.

A portion of the centre line setting-out has been shown in each case and needs little explanation.

The basis in Fig. 1 is the square and diagonal. Draw the centre line in each case with half the width of the bar on each side of it. The bars forming the outer edges of the panel will be half the full width of the bar plus half the nose band. No definite construction can be given for the cusps as the radii could be increased or decreased according to the size of the cusp and final effect required.

In Fig. 2 the basis is a series of equilateral triangles with the apex of each on the horizontal centre line of the panel. Two of the triangles are shown, ABC and DEF; the points of separation of the curves will be at the intersection of the sides of the triangles, these being common normals.

The square, circle, and quadrants of circles are the basis of Fig. 3. The diagonals of the square intersect at the centre of the circle and the angular points of the square are the centres for the quadrants.

The example in Fig. 4 is set out on a series of overlapping squares with their sides at 45° to the horizontal centre line. The centre of the circle is found by bisecting the angle formed by one side of the square and the horizontal.

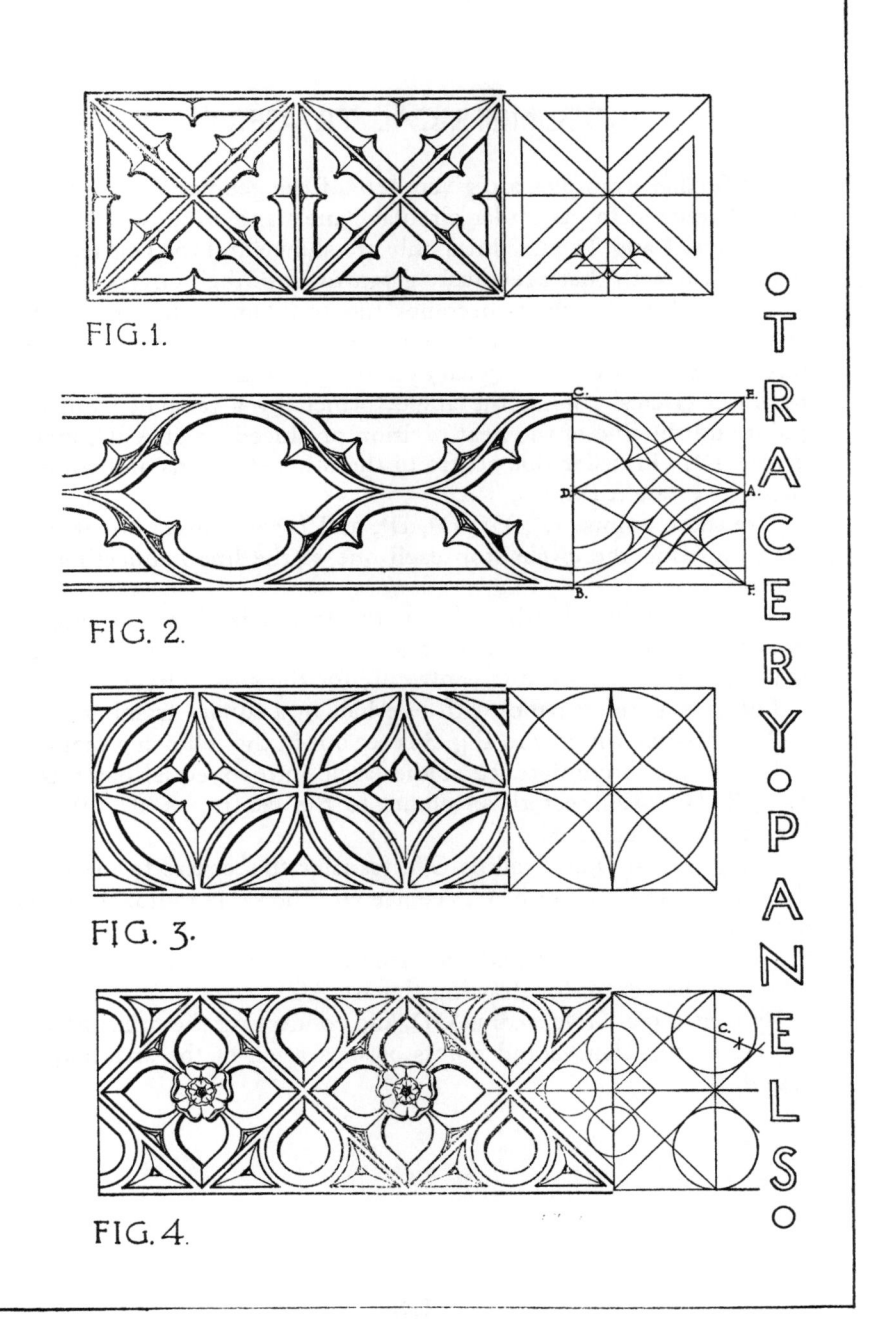

FIG. 1.

FIG. 2.

FIG. 3.

FIG. 4.

TRACERY PANELS

GEOMETRICAL TRACERY

SOME further examples are given here of the geometrical construction of tracery, in particular circular panels and windows.

In Figs. 1 and 2 centre lines only are shown and the construction is identical with that previously given for inscribed circles.

Taking Fig. 2, which becomes the problem of inscribing five circles within a given circle, commence by dividing the circumference of the circle into twice as many parts as the number of the required circles. Bisect the angle formed by a tangent from one point and the line of the next division produced, as at ABC, giving centre D^1. Transfer this centre to the other divisions and connect them by straight lines.

Use these points, D^1, D^2, D^3, D^4, and D^5 as centres but instead of completing the circle stop each one at the line connecting the centres.

A variation based primarily on the trefoil is shown in Fig. 4. A locus of centres between the main circle and the inscribed circle is used to find the striking centre, C, for the small circle.

Fig. 5 is a trefoil panel with one-half in a square framing.

The circular window in Fig. 3 is set out in the following manner.

Divide the circumference of the circle into six equal parts as at 1, 2, 3, etc. Connect these points to the centre, as at 2–0, and bisect giving centre A.

Using centre O and radius OA, transfer this centre to the other divisions B and C. With A as centre and the same radius draw the arc 2–C.

Join centres B and A and produce until it cuts the arc 2–C in D, which is the centre for drawing the arc AC.

By drawing a circle, radius OD, and connecting the other points, the remaining centres and points of separation of the arcs may be found.

GEOMETRICAL TRACERY

FIG. 1.

FIG. 2.

FIG. 3.

FIG. 4.

FIG. 5.

DEVELOPMENTS OF SPHERE

THE surface of a sphere cannot be developed accurately but there are methods by which an approximation may be obtained of sufficient accuracy for practical purposes. Two constructions are illustrated here, the drawings being arranged for convenience with two separate elevations above one plan.

The first method is by means of zones, and Fig. 1 shows diagrammatically the geometrical principle involved. A *zone* is a strip between two horizontal parallels.

If the sides of the zone are considered as straight lines or the generators of a cone, the problem becomes one of developing part of the surface of a cone having as base the bottom line of the zone.

In Fig. 2 half the surface of two separate zones has been developed. Divide the circumference of the elevation or section of the sphere into a number of parts, as at 1, 2, etc., in Fig. 2. These would correspond with the bed joints in the case of a stone dome.

Join the two division points of the zone to be developed by a straight line and produce until it meets the centre line as at 1–2 and C. C–1 is now the generator of the cone, the surface of which may be developed as previously described. The inaccuracy of the method lies in the fact that the straight line 1–2 is not the true length on the arc.

The second method, also an approximation, is by *gores* or *lunes*, and is illustrated in Figs. 3 and 4.

Divide the elevation of the sphere into a number of parts by means of horizontal lines and project the plan of the lines. By means of radial lines divide the plan of the sphere into a number of gores.

Produce the centre line of the gore, making its length equal to the length of the curve in section. Using 4 as centre draw a series of arcs from the division points and project on to them the length from plan.

The elevation of one of the gores has been projected in Fig. 3, but this was not necessary in order to obtain the development.

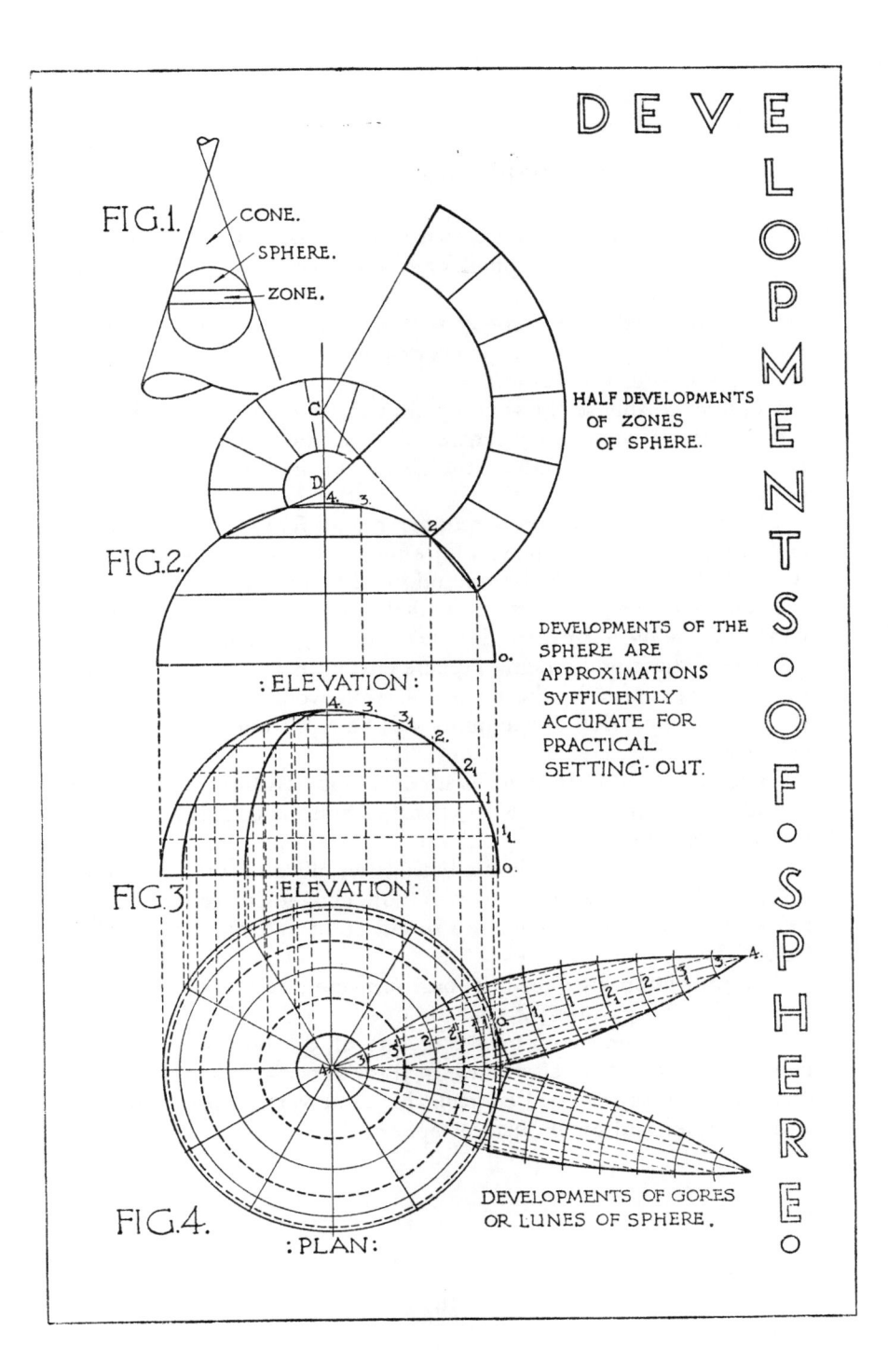

DEVELOPMENTS OF SPHERE

FIG.1. CONE. SPHERE. ZONE.

HALF DEVELOPMENTS OF ZONES OF SPHERE.

FIG.2.

:ELEVATION:

DEVELOPMENTS OF THE SPHERE ARE APPROXIMATIONS SUFFICIENTLY ACCURATE FOR PRACTICAL SETTING·OUT.

FIG.3 :ELEVATION:

FIG.4. :PLAN:

DEVELOPMENTS OF GORES OR LUNES OF SPHERE.

HEMISPHERICAL DOME

A DOME in its simplest form consists of half a sphere, and geometrically is a practical application of the exercise on plane cuttings through a sphere.

Constructionally the dome consists of a series of courses, the bed joints of which are conical, each course being divided into a number of blocks having vertical joints.

Commence the setting-out by drawing the section making the inner curve a semicircle, centre C^1, and the outer curve struck from C^2 in order to reduce the thickness and the weight at the crown.

Divide the section into a number of, preferably equal, courses and draw their horizontal arris lines in inside and outside elevation. A quarter each of the reflected and outside plans is sufficient.

Each course should next be divided into a number of blocks, the joint lines being normal in plan, i.e. radiating to the centre.

Each of these joints being a plane cutting is a portion of a circle in true shape, that is, the normal section would apply to each of them. The elevation of the joints will be a portion of an ellipse in each case.

Considering X–O in plan as a continuous cutting, its elevation would be the semi-ellipse with C'–O' as half major axis and C'–X' as half minor axis.

The elevation of all the joints both externally and internally could be found by this method, but it is more convenient to take one or more additional horizontal lines in elevation and plan and project from these.

The projection lines for one joint projected by this method have been shown by points 1, 2, and 3.

A sketch is shown in Fig. 2 of one of the blocks in the first course. The bottom bed surface in this course will be a horizontal plane surface, the top bed conical, the joints vertical plane surfaces and the inner and outer surfaces spherical.

○ HEMISPHERICAL ○ DOME ○

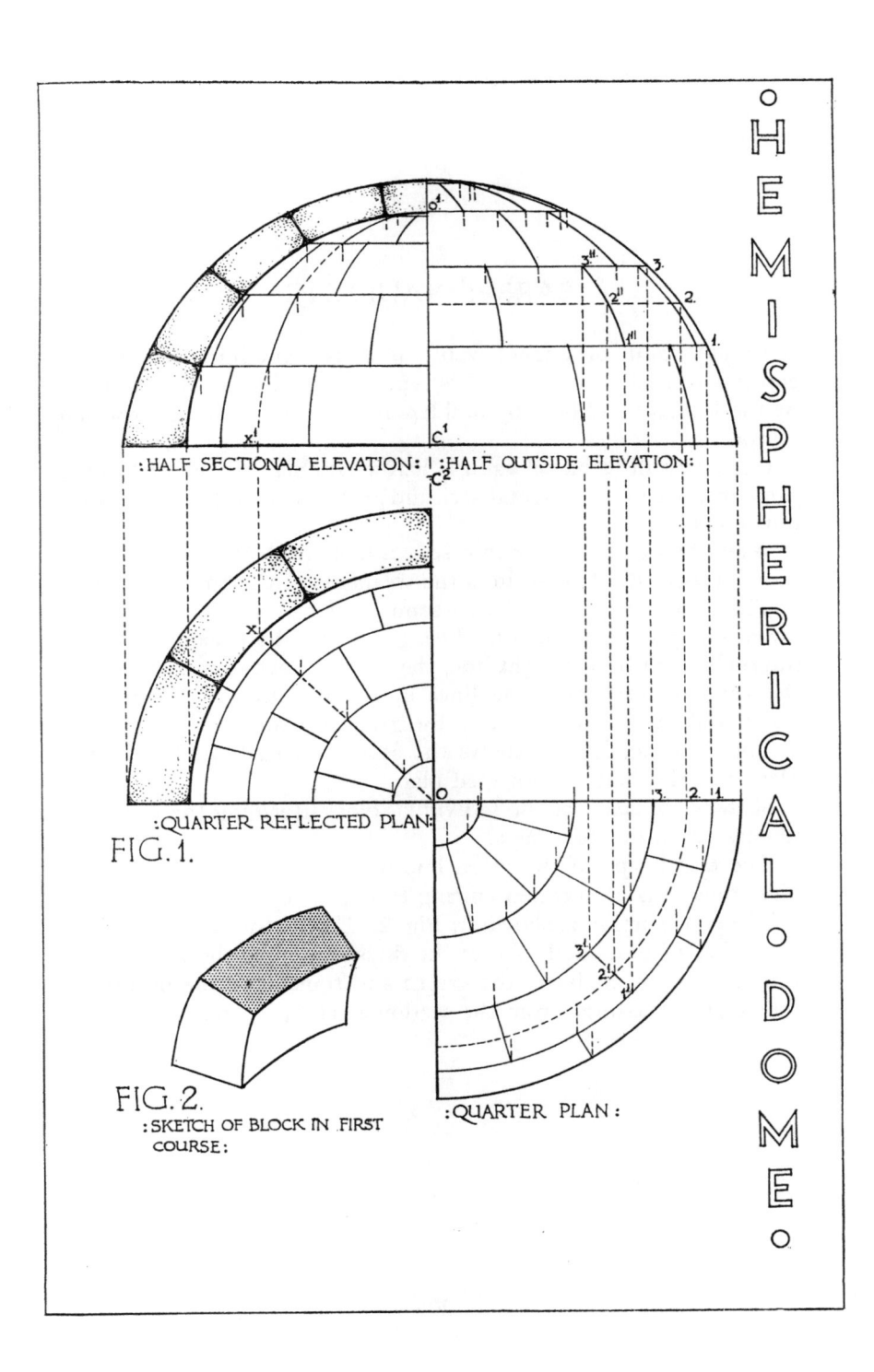

:HALF SECTIONAL ELEVATION: | :HALF OUTSIDE ELEVATION:

:QUARTER REFLECTED PLAN:

FIG. 1.

FIG. 2.
:SKETCH OF BLOCK IN FIRST COURSE:

:QUARTER PLAN:

BARREL VAULTING

A BARREL vault, or waggon vault as it is sometimes called, is a continuous arch or tunnel. This type of arched roof was first used by the Romans and later by builders in this country in the Norman period.

Geometrically a barrel vault is half a cylinder, it is jointed in the same manner as a segmental arch and presents no particular setting-out difficulty.

When two barrel vaults intersect, a line of intersection termed a groin is formed. The groin is the weakness of the construction.

The first example is of two semicircular vaults of equal span intersecting at right angles. The groin line, both externally and internally, will be a straight line, the diagonal of the square plan of the vaulted chamber. The lines of the radiating bed joints are horizontal and will intersect on the groin in plan.

The drawing, Fig. 1, shows the section of each vault, half the plan, and also half the reflected plan.

Sketches of some of the individual blocks forming the vault are illustrated on the next page.

The true shape of the groin line will be a semi-ellipse in each case, since it is an inclined cutting through a cylinder.

The construction is shown in Fig. 2. Divide the arch curve into a number of parts and project the divisions on to the groin lines. Project at right angles to the groins and transfer to the projectors the heights measured from the section as at A, B, etc.

BARREL O VAULTING

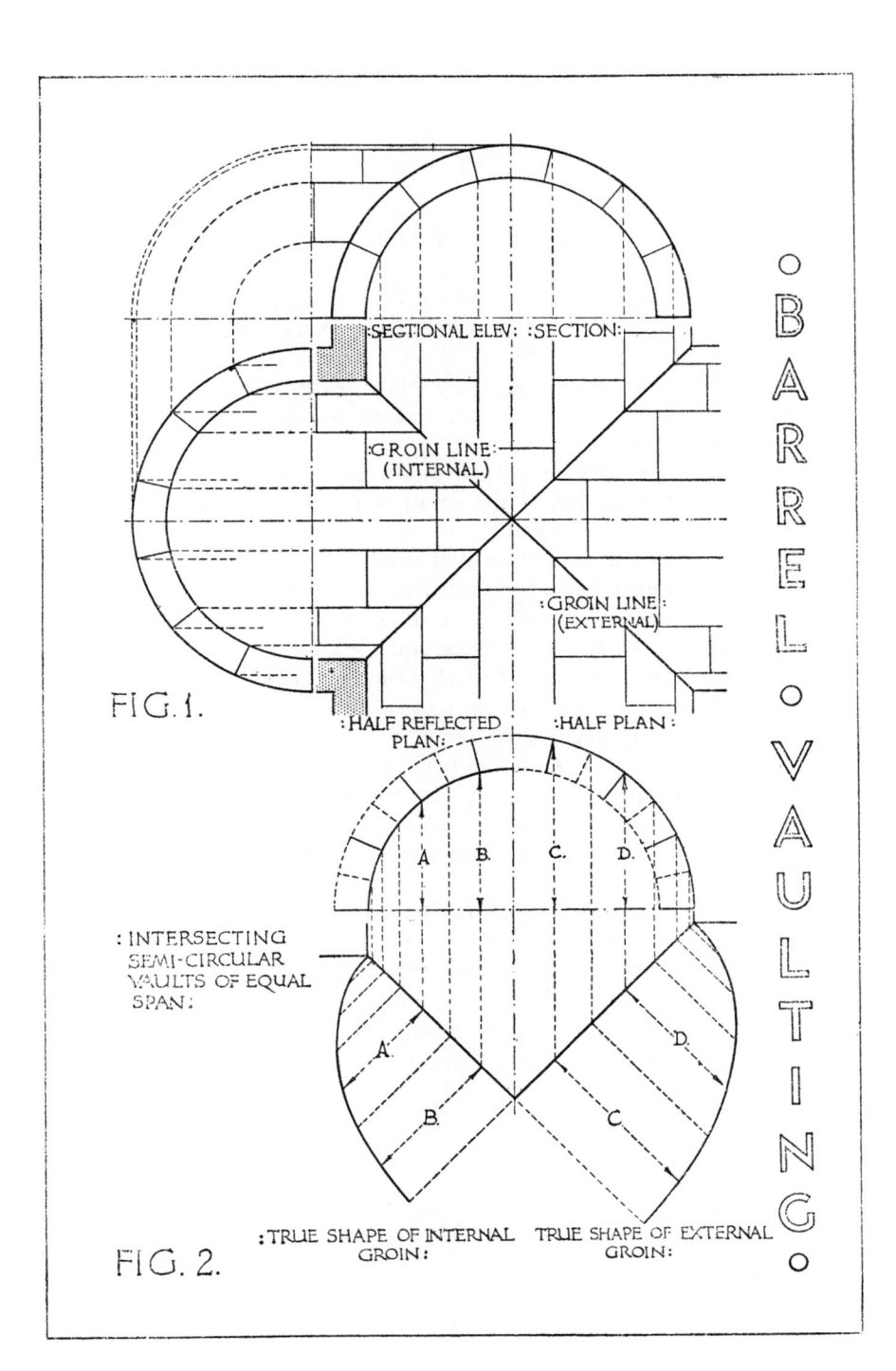

FIG. 1.

- SECTIONAL ELEV: :SECTION:
- GROIN LINE (INTERNAL)
- GROIN LINE (EXTERNAL)
- HALF REFLECTED PLAN:
- HALF PLAN:

FIG. 2.

- INTERSECTING SEMI-CIRCULAR VAULTS OF EQUAL SPAN:
- TRUE SHAPE OF INTERNAL GROIN:
- TRUE SHAPE OF EXTERNAL GROIN:

BARREL VAULTING

As a second example of barrel vaulting, Fig. 1 shows two vaults intersecting at right angles with their spans unequal but their heights remaining the same.

The groins will be straight lines both externally and internally, but they will not fall vertically one above the other if the wall thickness remains equal.

Suppose that, given the plan of the rectangular chamber to be vaulted and the section of the small vault, it is required to obtain the section of the large vault.

Divide the section of the small vault into a suitable number of stones and project on to the groin the plan of the joint lines. From the groin project in the direction of the intersecting vault. Insert a base line, X–Y, at right angles and measuring heights from the small vault transfer them to the larger section, as at A–A', B–B', etc. The large vault section will be a semi-ellipse.

It should be noted that although it is an advantage for the bed joints to be continuous horizontal lines both internally and externally in both vaults, this is only possible when there is no considerable difference in spans. If the difference is great the joints in the large vault are too far from normal to the curve to be constructionally sound.

The courses have been jointed in their length, and each of these joints will be a vertical plane surface.

Sketches of three of the stones each containing a portion of the groin line are shown in Figs. 2, 3, and 4. A springing stone, No. 1, is illustrated in Fig. 3; it is standing on a horizontal bed. Stones Nos. 2 and 3 are shown with their curved soffit surface uppermost.

A. B.

:SECTION SMALL VAULT:
(SEMI-CIRCULAR)

:SECTION LARGE :HALF REFLECTED :HALF PLAN:
VAULT:(SEMI-ELLIPSE)   PLAN:

FIG. 1.

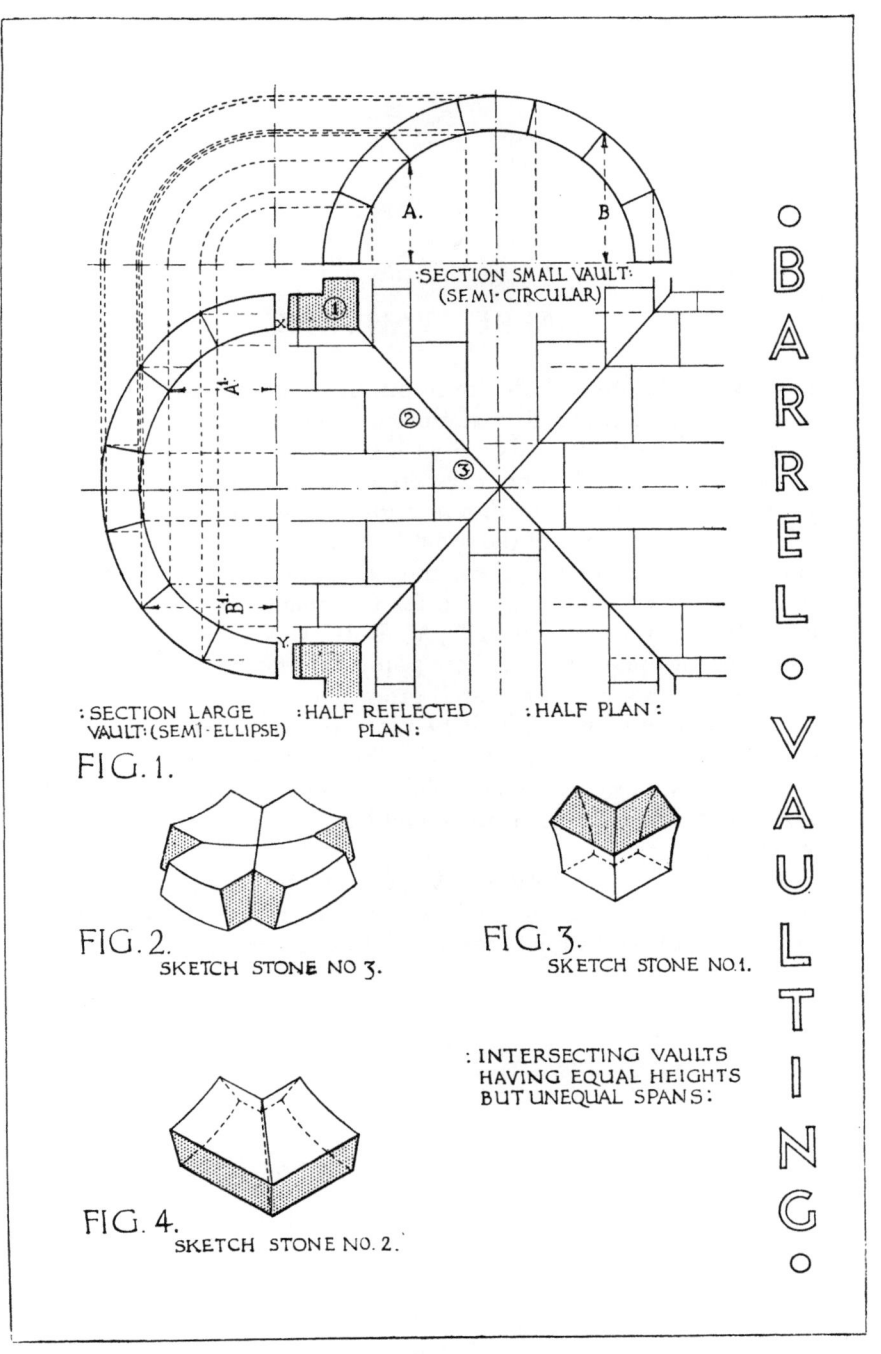

FIG. 2.
SKETCH STONE NO 3.

FIG. 3.
SKETCH STONE NO. 1.

:INTERSECTING VAULTS
HAVING EQUAL HEIGHTS
BUT UNEQUAL SPANS:

FIG. 4.
SKETCH STONE NO. 2.

○ BARREL ○ VAULTING ○

BARREL VAULTING

THE two vaults given here are in outline only and are intended to show further examples of variation in groin lines.

In Fig. 1 the vaults are at right angles and are of unequal span. To maintain a level height at the crown of the soffit the small vault has been stilted which results in a groin line curved on plan.

To obtain the shape of the groin line draw the sections of the two vaults and divide one of them into a number of equal parts. By rebatment or measurement transfer the heights of these divisions on to the other section, as at A–A', B–B', C–C'.

Project into plan each of these points, giving A", B", C", etc., as a series of points in the required groin line.

The example in Fig. 2 is of two intersecting semicircular vaults of unequal span and unequal rise. The groin lines will not meet.

Find the highest point at which the small vault intersects the large vault which in plan gives point A" as the extremity of the groin line.

Divide the height between A' and the springing of the large vault into a number of parts. Transfer the heights of the points to the small vault and project the plan of each as at A", B", C", etc.

BARREL ○ VAULTING

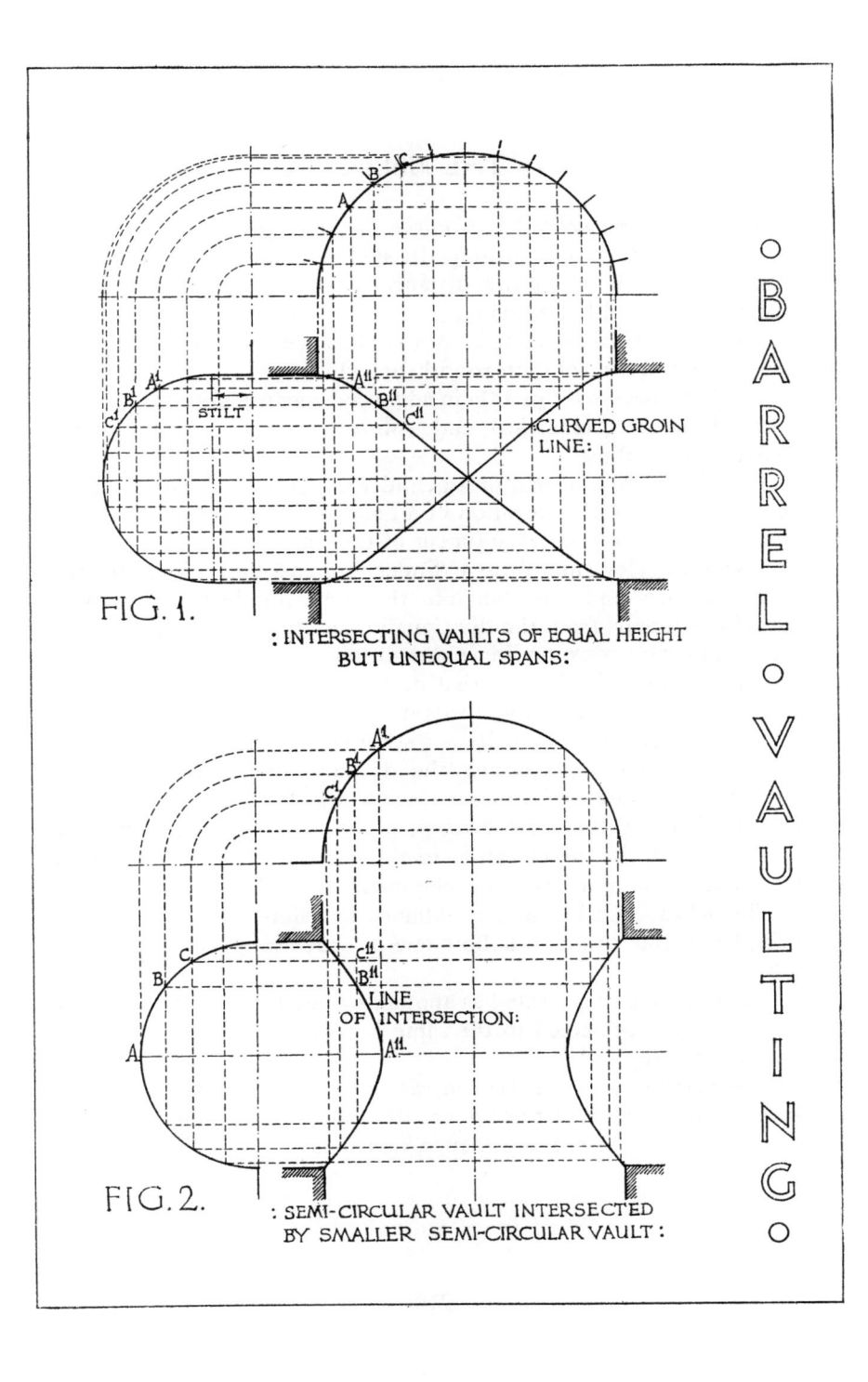

FIG. 1.

STILT

CURVED GROIN LINE:

: INTERSECTING VAULTS OF EQUAL HEIGHT
BUT UNEQUAL SPANS :

FIG. 2.

LINE OF INTERSECTION:

: SEMI-CIRCULAR VAULT INTERSECTED
BY SMALLER SEMI-CIRCULAR VAULT :

THE HELIX

The helix (plural *helices*) is a curve formed by a point travelling round the surface of a cylinder in such a manner that there is a constant ratio between the distance travelled vertically and the distance travelled horizontally.

The *axial pitch* of a helix is the distance travelled along the axis during one revolution round the cylinder.

The helix occurs frequently in advanced constructional problems, particularly in geometrical and spiral centre newel stairs and twisted handrailing.

The elevation and plan of a cylinder are shown in Fig. 1. It is required to draw the elevation of a helical line passing once round the cylinder, also the development of the line.

Divide the elevation of the cylinder into any convenient number of equal parts and the plan into the same number of divisions. Draw horizontals from the divisions in elevation and project vertically from the corresponding points in plan.

If the cylinder were solid that part of the elevation curve above point 7 would be hidden and therefore dotted.

The development of a helix is a straight line, and is the diagonal of the rectangle representing the development of the cylinder. Verify this by developing the surface of the cylinder as previously explained, using the same division points from the plan. Project horizontally from the elevation of the curve to vertical projectors from the division points in development.

The helical solid in Fig. 2 is obtained in a similar manner but the cylinder in this case is a hollow one in order to give width to the solid.

The inner curve is treated as another cylinder, the inner divisions on plan being projected to the same horizontal divisions in elevation, as at 6′–6″ and 8′–8″.

The heights are measured on each vertical division making the lower lines parallel to the upper. In Fig. 2 this height has been made equal to one of the elevation divisions.

HELIX

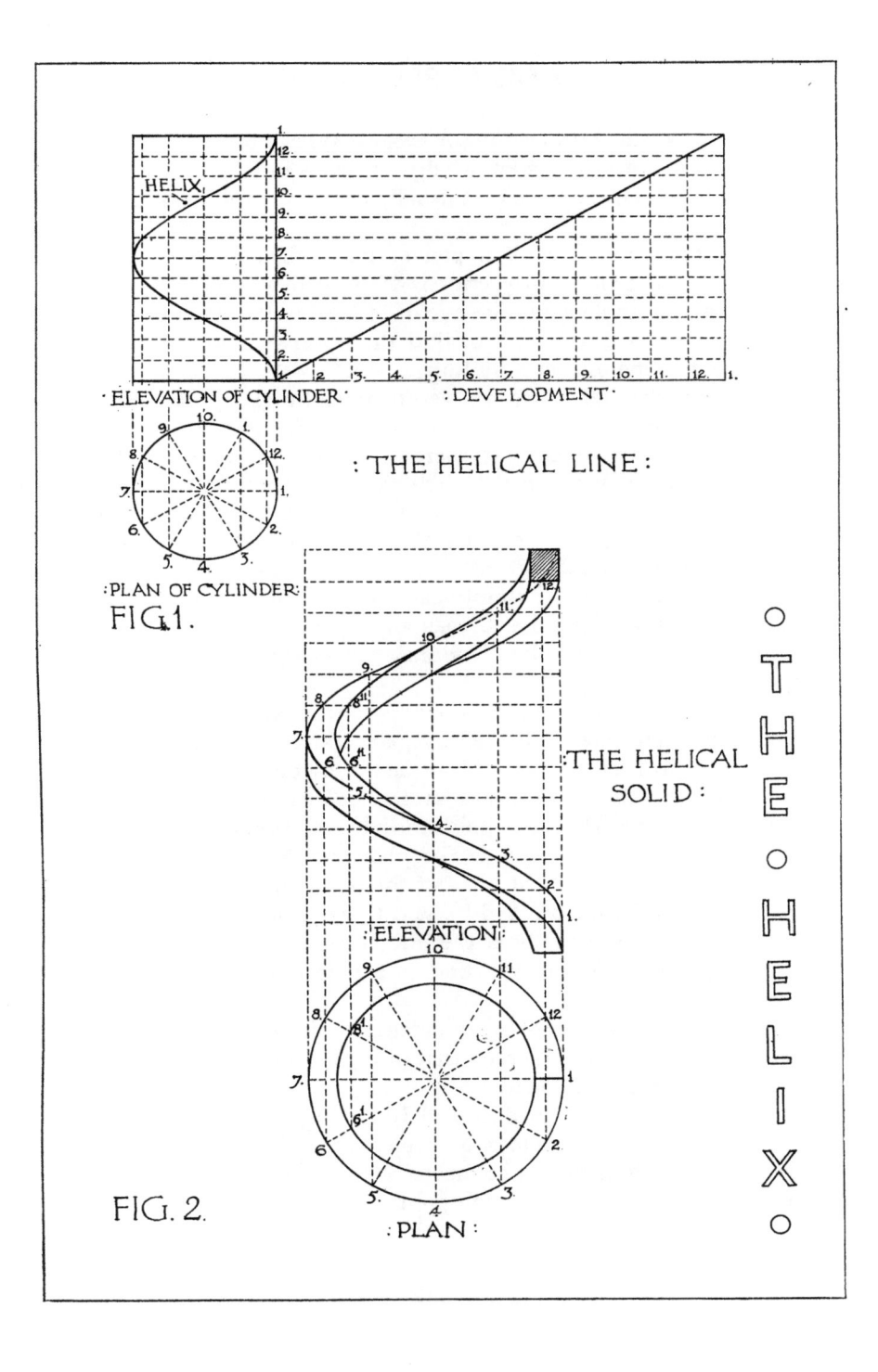

·ELEVATION OF CYLINDER·   :DEVELOPMENT·

: THE HELICAL LINE :

:PLAN OF CYLINDER:
FIG. 1.

THE HELICAL SOLID :

:ELEVATION:

: PLAN :

FIG. 2.

o THE o HELIX o

SHADOW PROJECTION

THE art of shadow projection (or *sciography*) can be a very advanced form of geometry, but a knowledge of its essentials is adequate for representing the shadows that usually occur on architectural elevations.

For architectural work a conventional shadow angle is used which in orthographic projection is 45° in plan and 45° in elevation. Actually, this angle would be one of 35° 16′ to the horizontal plane if the sun were above and behind the observer's left shoulder as he looked at the face of a building.

Further, the light rays are assumed to be always at the same angle, to come always from the left-hand side, and to be parallel.

For a shadow to be cast there must be one or more shadow planes onto which it can fall. The vertical plane, now termed the vertical shadow plane, is usually the face of the building, and the horizontal plane, now termed the horizontal shadow plane, is usually the ground.

For all but the most simple cases there will be more than one vertical shadow plane, i.e. breaks and openings in the face of the building, also the shadow plane may not be actually vertical, as in the case of roof slopes.

The best approach to the subject is to take examples of points in space and then lines, followed by planes and then solids.

A simple rule is to project from the given point at 45° in elevation and plan; whichever projector meets the shadow plane first, project up or down from that point to meet the other projector.

In Figs. 1, 2, and 3, A is the plan and A′ the elevation of a point in space. S is the shadow of the point and falls firstly on the H.S.P., secondly on the V.S.P., and thirdly, because the two projectors are equal, actually at the intersection of the two planes.

In Figs. 4, 5, 6, and 7 a line or a pole is treated in a similar manner.

The shadow is found by projecting each end of the pole as a separate point. It will be seen that in Figs. 4 and 7 the shadow falls partly in each of the planes.

The examples in Figs. 8, 9, and 10 are of planes in various positions relative to the planes of projection. Here each should be considered as four points connected by straight lines, the shadows being found separately of each point.

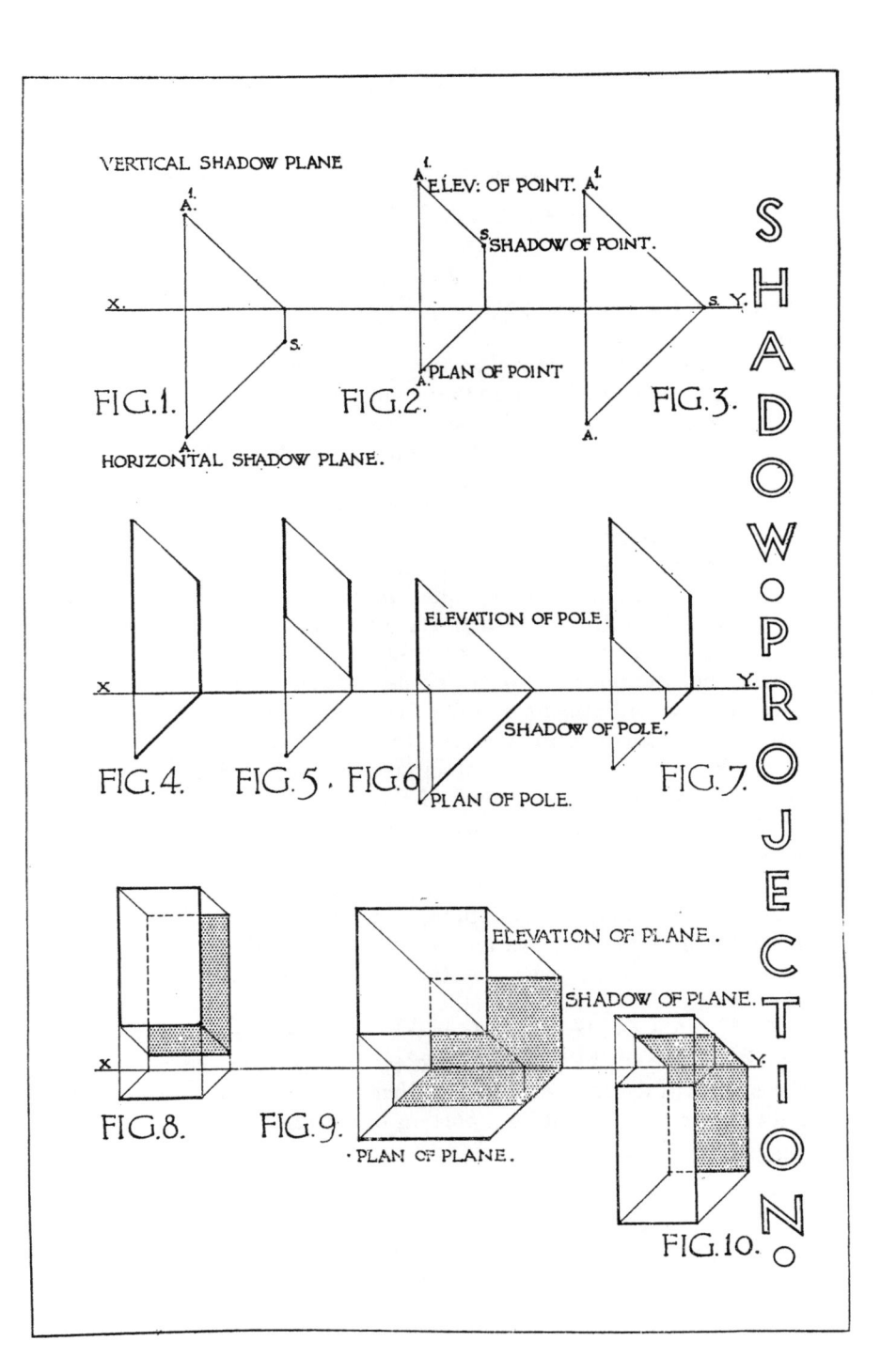

VERTICAL SHADOW PLANE

ELEV: OF POINT.
SHADOW OF POINT.
PLAN OF POINT

FIG.1.   FIG.2.   FIG.3.

HORIZONTAL SHADOW PLANE.

ELEVATION OF POLE.
SHADOW OF POLE.
PLAN OF POLE.

FIG.4.   FIG.5.   FIG.6.   FIG.7.

ELEVATION OF PLANE.
SHADOW OF PLANE.

FIG.8.   FIG.9.
PLAN OF PLANE.
FIG.10.

SHADOW PROJECTION

SHADOW PROJECTION

CONTINUING examples of shadow projection in the sequence suggested on the previous page, Figs. 1 and 2 show two square solids suspended in space.

These should be treated as two planes or four separate vertical lines, the shadow of the points forming the extremities of the lines, or the angular points of the planes being found separately.

In Fig. 1 part of the shadow will fall on the horizontal shadow plane but in Fig. 2 all the shadow is on the vertical shadow plane. In both cases part of the shadow will be hidden behind the actual figure—it is shown dotted.

The circle as a plane and a solid is shown in Figs. 3 and 4. In both cases the figure is suspended in space above the H.S.P. and in front of the V.S.P. The diameter of the circle and its centre have been projected as individual points. If a series of points is taken round the circumference of the circle and each projected, the resultant shadow will also be a circle. It is only necessary therefore to project the centre point to obtain the full shadow.

In Fig. 4 the centres of both circles are projected and connected by tangential lines, the shadow of the sides of the cylinder.

The circle in Fig. 5 is a plane in contact with the V.S.P., above and parallel to the H.S.P.

The resultant shadow is an ellipse on the vertical shadow plane. The plan is divided into a number of parts and 45° projectors taken from these points in elevation and plan. Where the projectors in plan meet the X–Y, project vertically to the corresponding lines in elevation.

In Fig. 6 a similar example is taken, but as the position of the figure relative to the planes of projection has been altered, part of the shadow falls in each plane. That portion of the shadow that falls on the H.S.P. will be a portion of a circle.

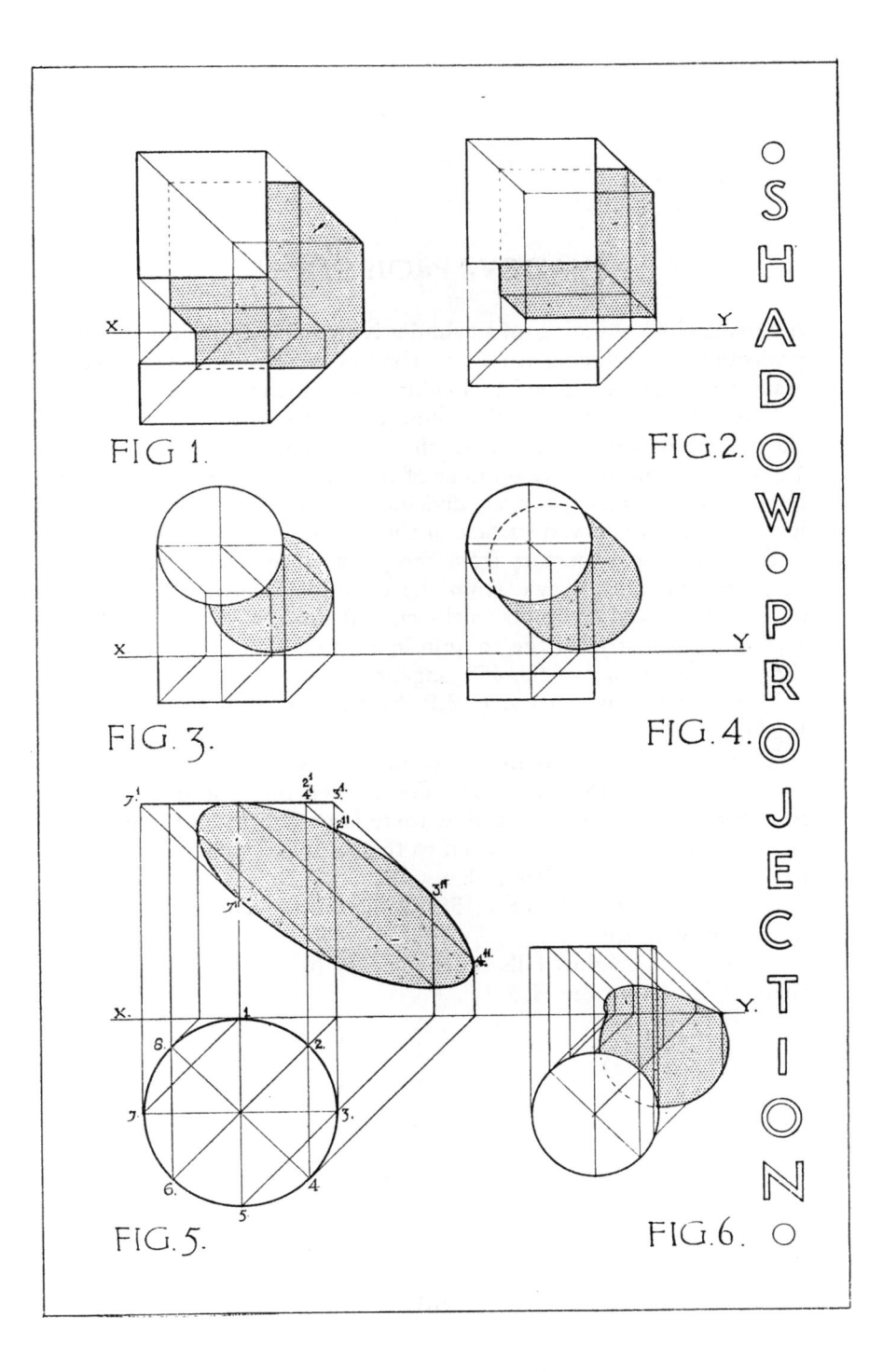

FIG 1.

FIG.2.

FIG.3.

FIG.4.

FIG.5.

FIG.6.

SHADOW PROJECTION

SHADOW PROJECTION

A SQUARE block standing on a cylinder is a good example for shadow projection, introducing a curved shadow plane. The two solids assembled may be likened to a column and square abacus.

The obtain the shade on the column itself and the shadow cast on the column by the abacus, divide that portion of the abacus between 1 and 4 on plan into any number of parts and project the divisions to elevation. Project from the divisions in plan at 45° until the projectors meet the curved surface of the column.

Draw similar projectors from the points in elevation until they meet vertical lines drawn from the column in plan. These give 0″, 1″, 2″, 3″, and 4″ points in the curved shadow.

The vertical strip of the column in shade is that portion beyond the point of contact of the 45° tangent.

The shadow cast on the H.S.P. by the solids is also shown in Fig. 1.

This could only occur in this form if the column were isolated, standing clear of the wall face. For convenience the ground line has been used as the X–Y and as there is no V.S.P. all the projectors from elevation will be taken to the X–Y and then up or down to meet the projectors from plan.

A similar solid is shown in Fig. 2, with one edge of the abacus in contact with the V.P.

Most of the shadow falls on the V.S.P. but part of the column shadow will be on the H.S.P.

SHADOW PROJECTION.

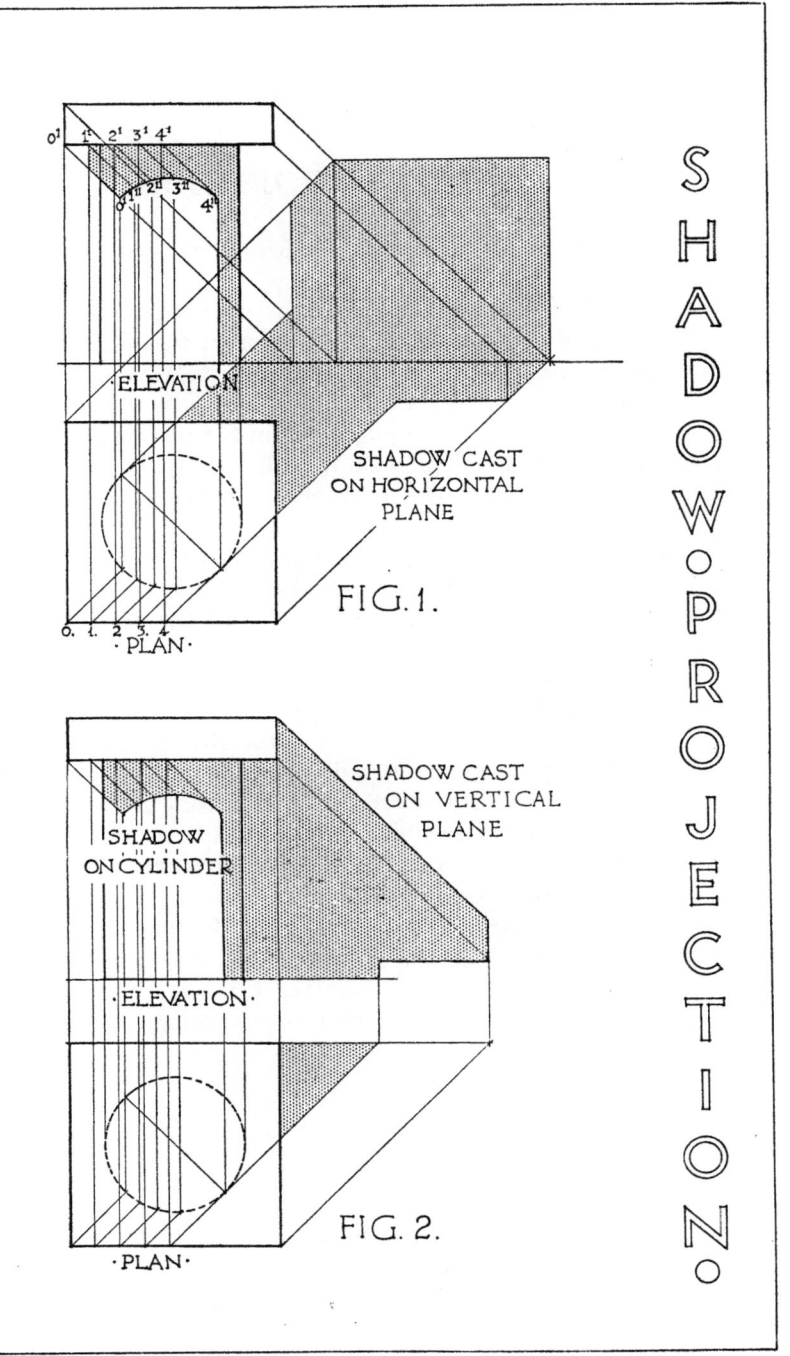

FIG. 1.

ELEVATION
SHADOW CAST ON HORIZONTAL PLANE
·PLAN·

FIG. 2.

SHADOW ON CYLINDER
SHADOW CAST ON VERTICAL PLANE
·ELEVATION·
·PLAN·

SHADOW PROJECTION

These examples show simple features that frequently occur in buildings.

In Fig. 1 a square recess or door opening is shown in plan, elevation, and section. It will be seen that the horizontal shadow line cast by the top edge of the opening could have been obtained either from the plan or the section.

Another recessed opening is shown in Fig. 2, this example having a semicircular head. The shadow will also appear as a portion of a semicircle and is obtained by projecting as a shadow point the striking centre of the arch.

The example in Fig. 3 is a projecting canopy over a recessed opening. A section has been given but has not been used to obtain the shadow, all the points in which have been projected from the plan.

A chimney-stack rising from a little above the eaves of a pitched roof is shown in front and side elevations, Fig. 4.

The sloping roof surface becomes the shadow plane onto which the points are projected from side elevation and then back onto the corresponding points projected from the front elevation.

The elevations of two dormer windows are shown in Figs. 5 and 6, together with side elevations to give the slope of the roof or position of the shadow plane.

The shadow cast on the elevation of the roof surface is found in both cases by projecting points at 45° from the side elevation onto the roof surface and projecting across to the corresponding points projected at 45° from the front elevation.

:RECESS:
FIG. 1

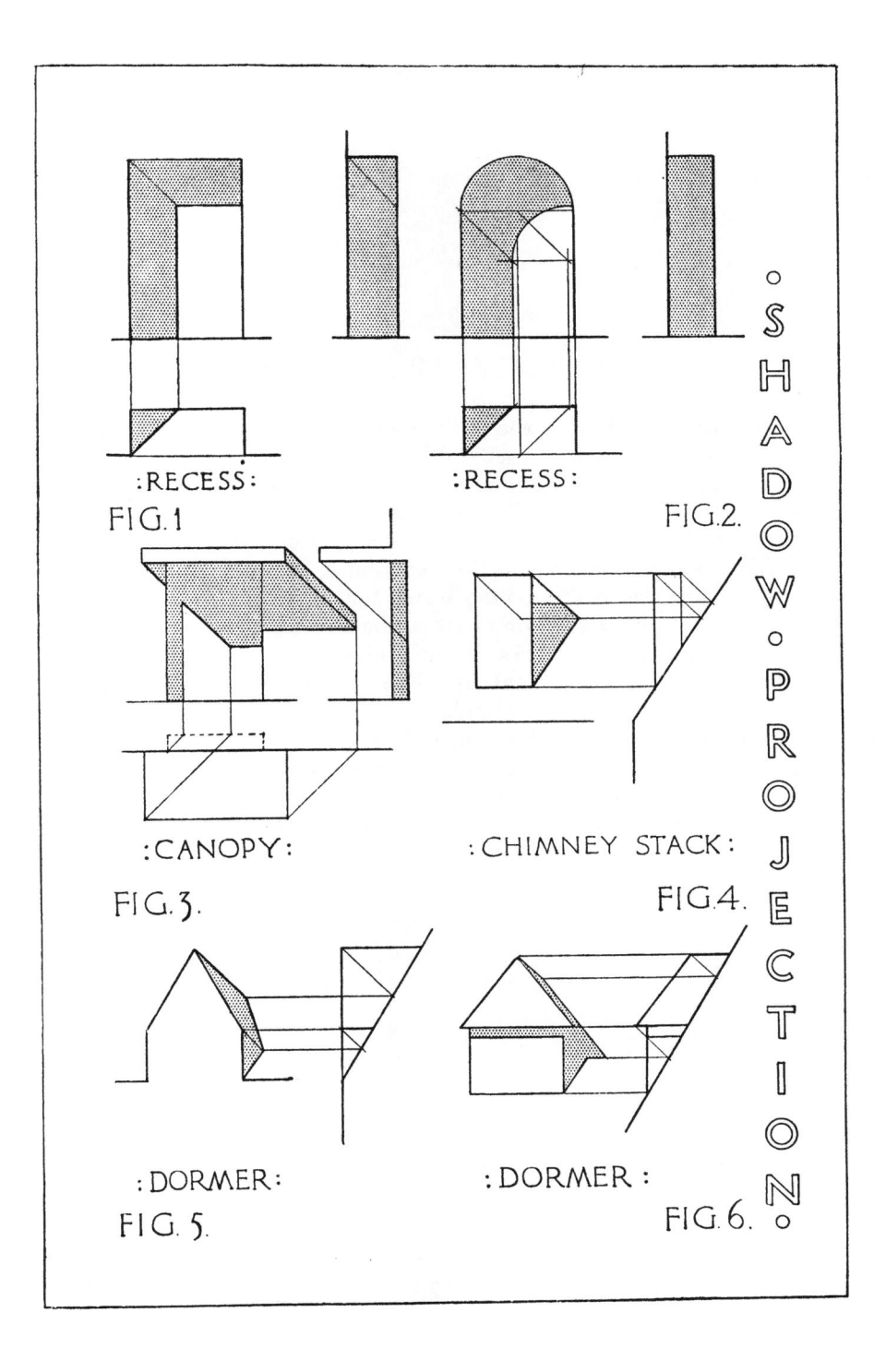

:RECESS:
FIG. 2

:CANOPY:
FIG. 3

:CHIMNEY STACK:
FIG. 4

:DORMER:
FIG. 5

:DORMER:
FIG. 6

SHADOW·PROJECTION·

SHADOW PROJECTION

THE area of shadow on the elevation of a moulded section is easily found from the section itself.

A series of 45° projection lines is drawn from the most prominent points of the moulding, and horizontal lines drawn from the points where these projectors touch the section again.

This is shown clearly on the elevation of the Doric pilaster cap, Fig. 1, and also on the entablature, Fig. 2. It will be noticed in Fig. 2 that the modillions cast a shadow on the metope space between the triglyphs. As the modillions occur on the return of the section this shadow can also be obtained in a similar manner and without reference to the plan.

The shadow cast by the return of the cap, Fig. 1, is found in the usual way projecting back on to the wall face which becomes the V.S.P. Where several of the members are together in shade the cast shadow will be a straight line on the V.S.P.

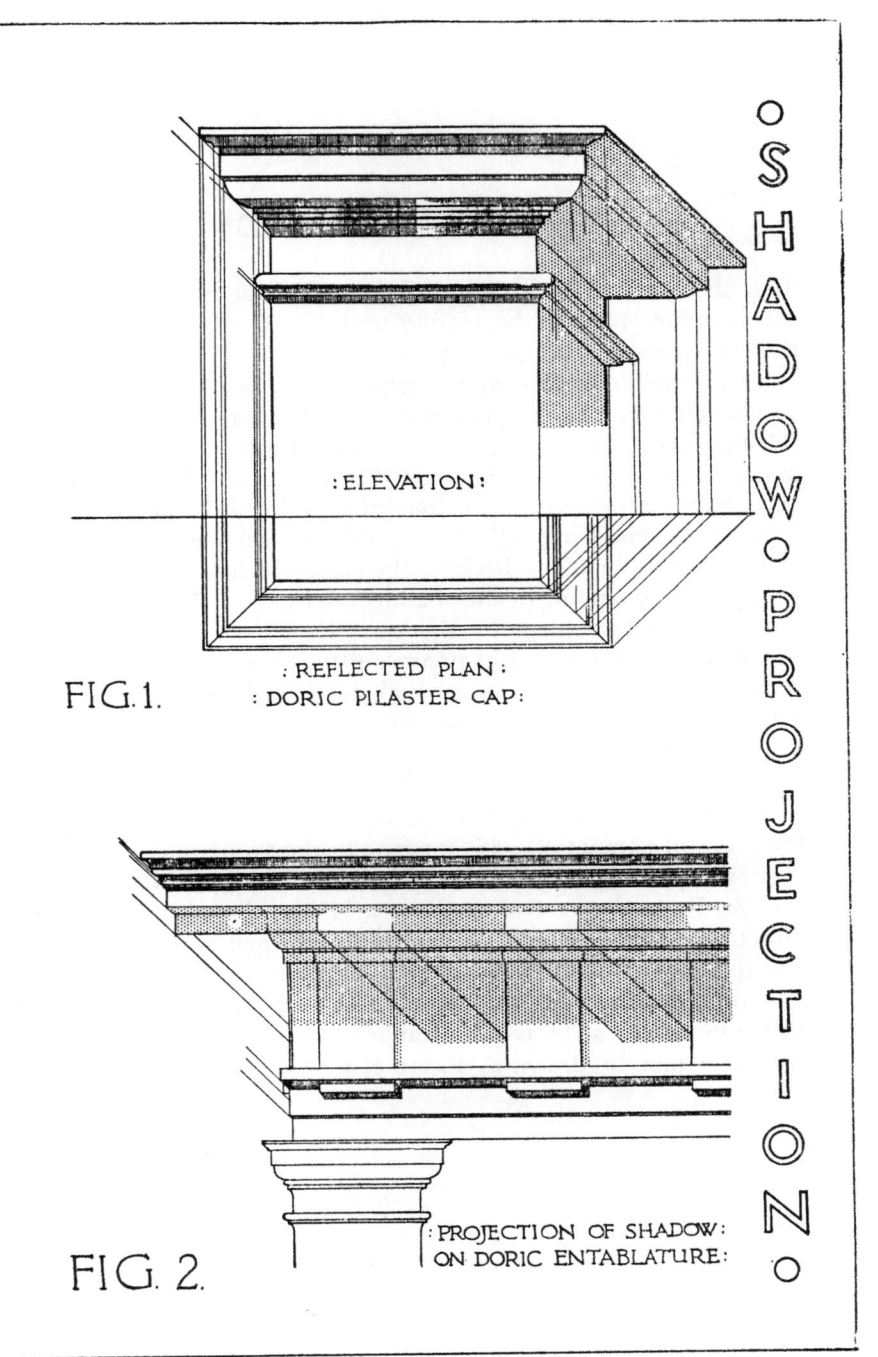

FIG. 1.
: ELEVATION :
: REFLECTED PLAN :
: DORIC PILASTER CAP :

FIG. 2.
: PROJECTION OF SHADOW :
: ON DORIC ENTABLATURE :

SHADOW PROJECTION

SHADOW PROJECTION

THE shadow cast in a niche with a spherical hood introduces a construction that will be readily understood after having studied the examples on sections of spheres.

The complete shadow and construction lines are shown in Fig. 1, and for clearness, with much of the construction omitted, in Fig. 2.

The shadow in the body of the niche is given by the line A–A″ in plan and is vertical as far as A‴ in elevation. Above this point and up to the springing of the hood the shadow is cast by the curved edge of the hood but still falls in the vertical body of the niche. B–B″ is a line in plan, B′–B‴ the same shadow line in elevation giving B‴ as a point in the shadow curve.

Above the springing the shadow falls on the spherical surface and is found by taking imaginary cuttings through the sphere.

Let C–C″ be the plan of one such cutting, the elevation of which is the quarter ellipse having as half major axis O–C′ and half minor axis O–C‴.

A shadow line drawn from C′ in elevation cuts the ellipse in C‴, a point in the required curve. The curve terminates at D, the point of contact of a 45° tangent to the semicircle.

A fair approximation to this curve may be obtained as shown in Fig. 3. From centre C draw lines at 45° to the springing in both directions. Divide one of these into three equal parts as at 1, 2, 3, and measure a fourth part of equal length beyond C as at 4.

A quarter ellipse drawn with CD as half major axis and C4 as half minor axis is the required approximation. The remainder of the curve may be drawn as an arc of a circle.

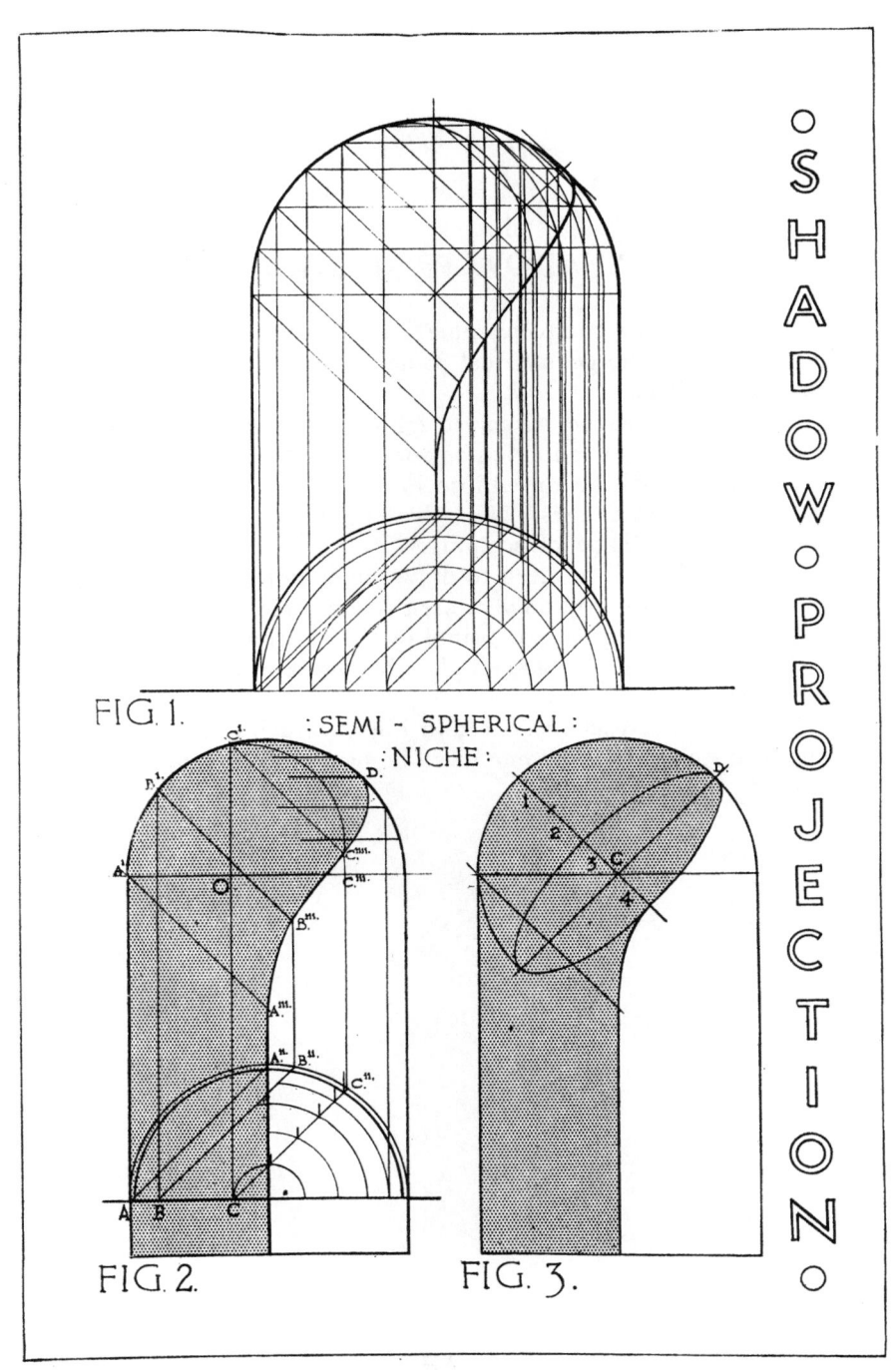

FIG. 1.

SHADOW PROJECTION

SEMI - SPHERICAL NICHE

FIG. 2.

FIG. 3.

SHADOW PROJECTION

THE shadows on a complete sphere, Fig. 1, are not often met with in practice but the geometrical principles involved apply to many general problems.

Again, as with the niche, imaginary cuttings or slicings are taken at 45°. Let A–B in plan be one such cutting plane, the elevation of which will be the ellipse having A'–O as half minor axis and C'–C' as major axis. Note the length C'–C' equals A–B because the true shape of section A–B is a circle having A–B as diameter. There is no need to draw the complete ellipse in elevation as the shadow will only be seen on one side. Tangents drawn at 45° to these ellipses are points in the shadow line which is in itself an ellipse. The plan of the shadow is also an ellipse projected from the elevation.

A similar construction is used in Fig. 2, which is a ring or could more probably be a torus mould. The plan of the shadow has been omitted.

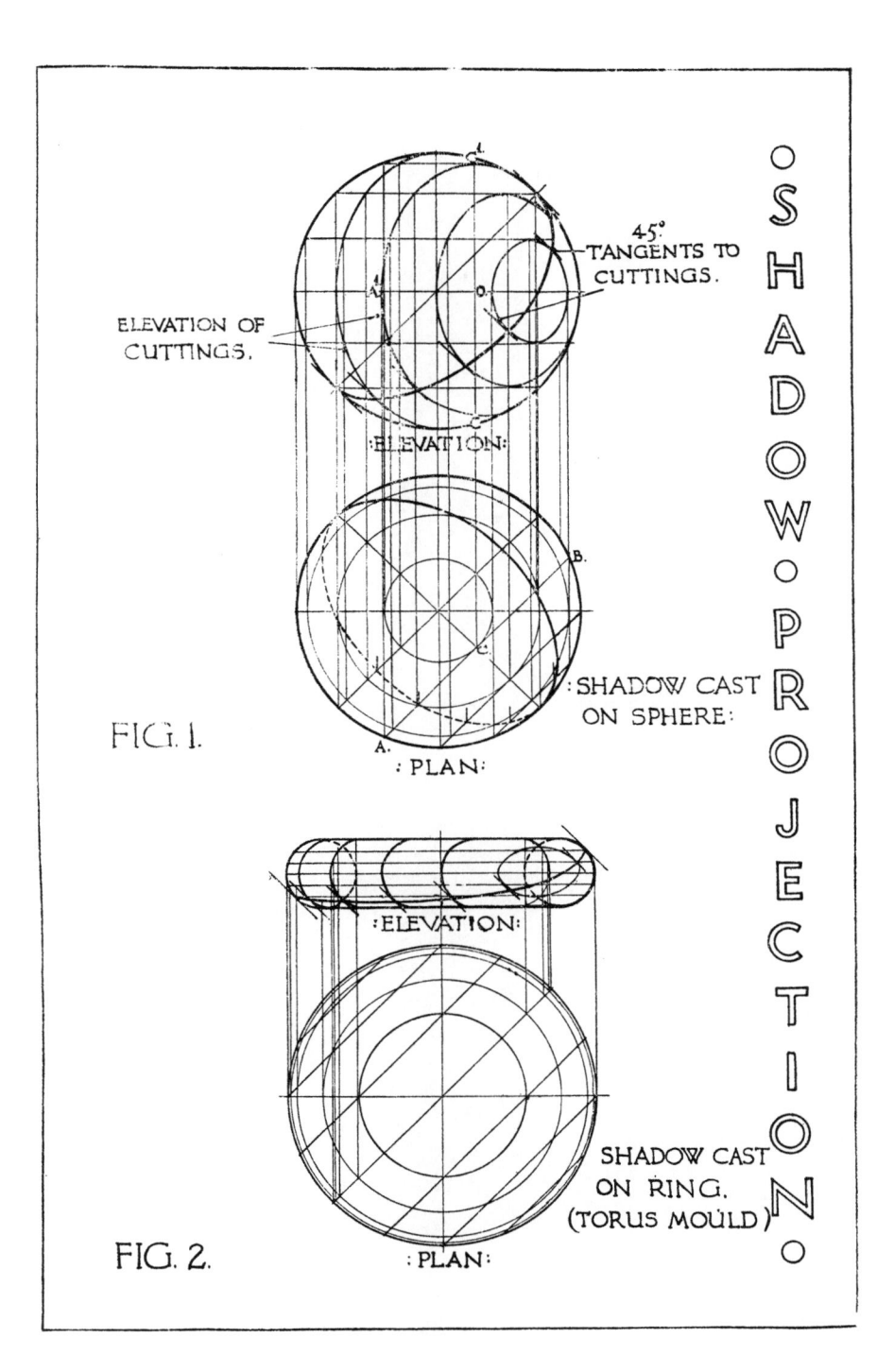

SHADOW PROJECTION

45° TANGENTS TO CUTTINGS.

ELEVATION OF CUTTINGS.

:ELEVATION:

SHADOW CAST ON SPHERE:

FIG. 1.

:PLAN:

:ELEVATION:

SHADOW CAST ON RING. (TORUS MOULD)

FIG. 2.

:PLAN:

INDEX

Angle, acute, 16
  bisection of, 14, 72
  by bisection, 16
  complementary, 18
  definition of, 12
  included, 20
  supplementary, 18
  types of, 16
Arc, definition of, 34
Annulus, area of, 108
Arches, 64, 66
Archimedean spiral, 116
Areas, calculation of, 108
  equivalent, 104, 106
Arrangement of drawings, 42
Astragal, 62
Auxiliary projection, 140, 144
Axes, of ellipse, 96
Axial pitch of helix, 210
Axonometric projection, 50, 58

Bar tracery, 194
Bead, 62
Bisection of angles, 14, 72
  of straight lines, 14
Blank tracery, 196
Bolection moulding, 84
Border, area of, 108

Canopy, 218
Cavetto, 62, 100
Change of ground line, 140
Chimney stack, 218
Chord, definition of, 34
Chords, scale of, 24
Circle, area of, 108
  definition of, 34
  in mouldings, 62
  inscribed, 78
  projections of, 52
  shadow projection of, 214
  tangents to, 70, 74
  touching given straight lines, 70, 72

Circumference of circle, 34, 108
Common normal, 74
  tangent, 74
Concoid of Nicomedes, 114
Cone, 122
  development of, 172
  projection of, 138
  sections of, 150, 164, 166
  surface area of, 110
  volume of, 110
Conic sections, 150–166
  in mouldings, 100
Constructions, geometrical, 12
Continuous curves, 82, 84
Cube, 122
  projections of, 124
  volume of, 110
Curved line, 12
Curves, arch, 64–6
  continuous, 82, 84
  spiral, 116–20
Cuttings of sphere, 168
  through moulded sections, 148
Cylinder, 122
  development of, 170
  oblique, 110
  projections of, 130
  surface area of, 110
  volume of, 110
Cyma recta, 62, 100
  reversa, 62, 100

Definitions of common terms, 12
Development of cylinder, 170
  of helix, 210
  of prism, 170
  of sphere, 200
Diagonal scales, 36
Diameter, 34
Directrix of parabola, 98
Dodecahedron, 13
Domes, hemispherical, 202
Dormer, 162, 218
Drawings, arrangement of, 42

INDEX

Drop arch, 66
Duodecagon in a circle, 30

Elevations, auxiliary, 140
   definition of, 40
Ellipse, approximations to, 102
   area of, 108
   as a conic section, 150
   as a plane figure, 96
Enlargement by squares, 38
Entasis of column, 114
Equilateral arch, 66
Equivalent areas, 104, 106
External tangent, 76

First angle projection, 40, 42
Focus of parabola, 98
Foil, 80
Foiled figures, 80
Fret patterns, 88
Frustum, 122
   volume of, 110

Geometrical constructions, 12
   solids, 122
   tracery, 198
Gibb's rule, 118
Goldman's rule, 120
Gores, 200
Gothic lettering, 8
   mouldings, 84
Guilloche, pattern, 92

Helix, 210
Hemispherical domes, 202
Hexafoil, construction of, 80
Hexagon in a circle, 30
Hexagonal prism, development of, 170
   projection of, 128
   sections of, 144
Hexagonal pyramid, projection of, 136
   sections of, 146
Hexahedron, 134
Horizontal section, 40
Horseshoe arch, 64
Hyperbola, 98, 166

Icosahedron, 134
Inscribed circles, 78
Internal tangent, 76

Interpenetration, 174–86
Intersecting mouldings, 188
Inverse similitude, 112
Ionic volute, 118–20
Irregular figures, enlargement of, 38
Isometric projection, 44, 46, 56

Jamb stop, 188

Key patterns, 88

Lancet arch, 64
Lettering, 8, 10
Lines, 12, 14, 154
   projections of, 154
   true length of, 158
Loci, 86
Logarithmic spiral, 116
Lunes, 186

Margin, area of, 108
Moorish arch, 64
Moulded sections, cuttings through, 148
   shadow projection of, 206
Mouldings, 68, 84, 100, 192
   circles in, 72
   Gothic, examples of, 84
   intersecting, 188
   pediment, 190

Niche, shadows cast in, 222
Normal, 34
   common, 74
   to an ellipse, 96

Oblique cylinder, volume of, 110
   projection, 48, 60
Octagon, construction of, 30
Octagonal pyramid, 146, 172
Octahedron, 134

Panels, tracery, 196
Parabola, 98, 164
Parallel lines, 12, 26
Parallelogram, 26, 108
Parallelopiped, 122
Patterns, 88–94

INDEX

Pediment mouldings, 190
Perimeter, 34
Perpendicular, construction of, 14
Perspective, 56
Pipes, intersection of, 176
Pitch, axial, of helix, 210
Plane, definition of, 12
Plane geometry, 12
Planes, oblique, 160
   projections of, 156
   shadow, 212
Plans, 40
   auxiliary, 140
Plate, tracery, 194
Point, definition of, 12
   of contact, 34
   projection of, 152
Polygons, 30
   area of, 108
   construction of, 32
Prism, 110, 122
   hexagonal, 128, 144
   square, 144
   triangular, 126
Projections, 40–60, 152–6
   arrangement of, 42
   auxiliary, 140, 144
   axonometric, 50, 58
   isometric, 44, 46, 56
   oblique, 48, 60
   of circle, 52
   of lines, 154
   of planes, 156
   of points, 152
   orthographic, 40
   pictorial, 44
   radical, 112
   shadow, 212–24
   solids, 124–38
   third angle, 42
Proportional scale, 38
Pyramid, definition of, 122
   development of, 172
   projection of, 132–6
   sections of, 146
   volume of, 110

Quadrant, definition of, 34
Quadrilateral, definition of, 26
Quatrefoil, construction of, 80

Radical projection, 112

Radius, definition of, 34
Raking sections, 192
Rampant arch, 64
Rebatment, 144
Recess, shadow projection of, 218
Rectangle, area of, 108
   construction of, 28
   definition of, 26
Reflected plan, 40
Representative fraction, 36
Rhombus, 26, 28
Right-angled triangles, 22
Ring, area of, 108
Roof surfaces, 162

Scales, construction of, 36, 38
Sciography, 212
Scotia, 62, 100
Sectional elevation, 40
Sections, 40
   of a cone, 166
   raking, 192
   solids, 142–8
Sector, 34
   area of, 108
Segment, 34
   area of, 108
Segmental arch, 34, 66
Semicircle, 34
Semicircular arch, 66
Shadow angle, conventional, 212
   planes, 212
   projection, 212–24
Similitude (similar figures), 112
Solid, definition of, 12
Solid geometry, 12
Solids, geometrical, 122
   projections of, 124–38
   sections, of 142–8
Specifications, standard, 42
Sphere, shadows on, 224
   cuttings of, 156
   definition of, 122
   developments of, 200
   surface area of, 110
   volume of, 110
Spiral curves, 116
Square, 12
   area of, 108
   construction of, 28, 106
Stilted arch, 64
Straight line, definition of, 12
Surface, definition of, 12

INDEX

Tangent, 34
  types of, 70, 74, 76
Tetrahedron, 134
Third angle projection, 42
Three-centred arch, 66
Tracery, 194–8
Trajan's column lettering, 10
Trammel method, 96
Transom, 188
Trapezium, 26, 28
Trapezoid, 26
  area of, 108
Trefoil, 80, 184

Triangle, area of, 108
  construction of, 20, 22, 106
  definition of, 12, 18
  types of, 18, 22
Tudor arch, 66
Tympanum, 190

Vaulting, barrel (waggon) 204–8
Volumes, calculation of, 110
Volute, ionic, 118, 120

Zone of sphere, 200

CPSIA information can be obtained
at www.ICGtesting.com
Printed in the USA
BVHW04*1141150718
521208BV00012B/110/P